History, Philosophy and Theory of the Life Sciences

Volume 18

More information about this series at http://www.springer.com/series/8916

Marta Bertolaso

Philosophy of Cancer

A Dynamic and Relational View

Springer

Marta Bertolaso
Institute of Philosophy of Scientific
and Technological Practice
University Campus Bio-Medico of Rome
Rome, Italy

ISSN 2211-1948 ISSN 2211-1956 (electronic)
History, Philosophy and Theory of the Life Sciences
ISBN 978-94-024-0863-8 ISBN 978-94-024-0865-2 (eBook)
DOI 10.1007/978-94-024-0865-2

Library of Congress Control Number: 2016943116

Printed on acid-free paper

This Springer imprint is published by Springer Nature
The registered company is Springer Science+Business Media B.V. Dordrecht

Contents

Introduction

According to Robert Weinberg (2014), "Those who have participated in cancer research during [the last 40 years] have witnessed wild fluctuations from times where endless inexplicable phenomenology reigned supreme to periods of reductionist triumphalism and, in recent years, to a move back to confronting the endless complexity of this disease" (p. 267). Much empirical information on cancer has been produced, and many theoretical models have been formulated to make sense of this proliferating body of research.

To a first approximation, conflicts in the interpretations of cancer take place on the battleground between reductionism and anti-reductionism. This challenge has been well summarized by Dupré:

> Certainly, no one knows how to explain all the properties of a complex organism in terms of the properties and arrangements of its parts; the question is whether this is simply a reflection of the underdeveloped state of our current biology, or whether there are deeper obstacles, obstacles in principle, that will continue to prevent us from doing this (Dupré 2010, 34).

Since the 1970s, the origin of cancer is being explored from the point of view of the Somatic Mutation Theory (SMT), called this way by critics Ana Soto and Carlos Sonnenschein (e.g. 1999). The SMT focuses on genetic mutations and clonal expansion of somatic cells. As cancer research expanded in several directions, the dominant focus on cells set up by the SMT remained steady, but the studies multiplied the classes of genes and the kinds of extra-genetic factors that were shown to have causal relevance in the onset of cancer. The wild heterogeneity of cancer-related mutations and phenotypes, along with the increasing complication of models, led to the oscillation described by Weinberg between, on one hand, the hectic search of "the" few key factors that cause cancer, and, on the other hand, the discouragement in face of a seeming "endless complexity".

To tame this complexity, cancer research started to avail itself of the tools that were being developed by Systems Biology. Networks, fields, attractors, collective dynamic behaviors and other methodological tools were put to work in analyzing the intricacies of relationships among huge numbers of factors. The continuity between knowledge and praxis that lies at the basis of the evolution of interpretative models of cancer highlights the relevance of epistemic virtues for a well-led scientific work. Such virtues imply major awareness for the assumptions, even philosophical ones, of any scientific activity and carry with them a lot of trust in human reason. Adhering to the experimental reality, human reason, in fact, is continuously exploring new explicative categories and choosing the most adequate for empirical purposes, but can also review these assumptions when logical coherence conflicts with new data and empirical evidence.

Anti-reductionist voices began claiming that cancer research was stuck in a sterile research paradigm. They pointed out that a cell-gene-centered perspective was at odds with several domains of knowledge, namely clinical practice and known cases of spontaneous or induced tumor reversion, and philosophical reflections on living beings and their peculiar dynamics. This alternative discourse even gave birth to a theory: the Tissue Organization Field Theory (TOFT). The latter defines itself anti-reductionist for assuming emergentism and organicism as default in accounting for carcinogenesis. Among the tenets of TOFT, there is the claim that the tissue is the right context in which the origin and establishment of the phenotype of tumor cells can be explained. Tissue-specific fields orient or dis-orient cells. Once compromised, they would cause cancer in a top-down way; conversely, the tissue can act as a tumor suppressor factor.

In a volume of *Nature Clinical Practice* of 2006, an article was published that captured my attention. The title was encouraging for the analysis I was trying to develop on cancer research and recent findings on cancer biology: "Dichotomies in cancer research: some suggestions for a new synthesis" (Sporn 2006). For Sporn, "Continuing high cancer incidence and mortality raise concern about the prevailing overall approach to the control of this disease." The author then elaborates on *fundamental dichotomies* between traditional and revisionist viewpoints (Table 1). Two of Sporn's seven dichotomies concern biological issues and call for a new synthesis in cancer research. Three dichotomies bear on methodological issues. Dichotomy 6 clearly refers to the reductionist-anti-reductionist debate, and dichotomy 7 seems methodological, but – I will argue – implies something much deeper about scientific practice and philosophical assumptions of cancer research.

Sporn then attempts a synthesis among these contrasting perspectives, considering topics such as the importance of controlling carcinogenesis in its earliest stages; the acknowledgement of epigenetic, as well as genetic, factors in cancer; the development of appropriate genetic animal models of carcinogenesis; the need for multifunctional agents to prevent and treat cancer; and the limits of reductionism. Another important topic of reflection is the "need for development of new preventive and therapeutic measures that will maintain quality of life, not merely extend life." Finally, Sporn highlights one aspect that will be central to the View developed in this book: the importance of context in cancer biology. Walt Whitman's famous quotation that "Nothing out of its place is good and nothing in its place is bad" (Sporn 2006, p. 364) is, indeed, one of my brightest guiding lights, and I will have more to say on how the terms in this phrase should be taken to mean.

Table 1 Dichotomies in cancer research according to Sporn (2006)

1.	'The disease is cancer' versus 'the disease is really carcinogenesis'
2.	'Cancer is a genetic disease' versus 'cancer is also an epigenetic disease'
3.	'Emphasis on cure of end-stage disease' versus 'prevention of early disease progression'
4.	'New emphasis on transgenic mouse models with single gene disruption' versus 'classical carcinogenesis models that damage multiple genes'
5.	'New emphasis on monofunctional agents' versus 'need for multifunctional agents'
6.	'Reductionism' versus 'the whole can be greater than the sum of its parts'
7.	'Hypothesis-driven research' versus 'the need for observational research'

What I do in this book is to demonstrate that a radical philosophical reflection is necessary to drive cancer research out of its empasses. At the very least, this will be a reflection on the *assumptions* of different approaches in cancer research, on the *implications* of what cancer research has been discovering over 40 years and more, on the *view* of scientific practice that is able to make more sense of the cognitive and social conflicts that are seen in the scientific community (and in its results), and, finally, on the *nature* of living entities with which we entertain this fascinating epistemological dance that we call scientific research.

As I shall show, the implications of a systemic approach, methodologically exemplified by Systems Biology, go well beyond the availability of technical tools to examine huge amounts of quantitative data on genes and other molecules. Systemic approaches lead to think of living organisms not only as organized molecular systems, but also as *organizers* of molecular systems. It does not only answer old questions in a new way: it drives questions over the dynamic maintenance of functional unity of biological entities. The whole is more than the sum of its parts, in so far as it has properties that are not encountered in the parts themselves, and that the parts are transformed once the whole has been integrated.

The TOFT issues a genuine challenge to cancer research. With new research protocols and reinterpretations of available empirical data, the TOFT not only demonstrates the crucial causal relevance of the tissue level but also introduces new concepts (e.g., "field"). The TOFT is indeed forceful in proposing itself as a radical philosophical alternative, with a rich use of a philosophical vocabulary (including, for example, reductionism, holism, emergence, etc.). TOFT authors criticize several features of the SMT, like the causal and explanatory relevance of somatic mutation for the origin of cancer and its onset. A deeper and more philosophical analysis, however, shows some limits in this anti-reductionist campaign. From a philosophical point of view, it has been acknowledged that when spatiotemporally continuous causal processes are at work no unique explanatory tool is either necessary or sufficient: "We need a theory of explanation that captures several different possibilities" (Woodward 2011). The conflict between SMT and TOFT makes up a debate over what is most fundamental in scientific explanation: *either* cells and genes *or* tissues. The debate needs to be brought back on more solid grounds: by resolving the conflict between different *ways of* explaining, we must understand better *what it is* that we are explaining. What is really at stake is not a privileged level of the biological organization at which the explanation of carcinogenesis should take place, but how the adequate observables are chosen in the process of scientific understanding of a complex phenomenon like cancer. We should keep looking for what is *really* fundamental in explanatory terms. Abandoning the idea that such "fundamentality" is something that can appear *within* our explanations or *inside* a natural entity, we should consider that "fundamental" is something that belongs to the process of scientific understanding and knowledge of the natural world through science. The world reacts and answers to our questions with the language it has, data. But as in any dialogue a semantic framework is required. Such framework requires a deeper understanding of scientific practice and of why science works. The explanatory import of this or that factor belongs to the interplay between the observable and the observer. The explanatory problems that anti-reductionist authors correctly raise must then be reframed.

The pathologic character of tumor cells goes beyond any genetic or biochemical alteration. How could cancer really be a matter of a proliferating cell? All cells do proliferate. In fact, as discussed with John Dupré and others, a more challenging question is why don't multicellulars develop cancer much more often. The neoplastic character of a cell is a matter of lack of integrated proliferation, which implies a meaning, a functional proliferative behavior that refers to the phenotype: by proliferating, neoplastic cells are not making up a phenotype. They are doing something else, by using the same metabolic pathways. We will talk about cancer in terms of a "natural history" which is inseparable from the "life history" of the organism. Tumor heterogeneity makes it apparent how different levels of organization are lost in the neoplastic process.

Besides cancer biology, the stakes seem to be a new understanding of how science works in practice when dealing with complex biological systems or multi-level biological processes. The most relevant outcome regards the possibility of a real pluralism of descriptive and explanatory account of complex biological systems, embracing the irreducibility of biological explanations both in epistemological and ontological terms.

Structure of the Book

I start by presenting the challenging biological features of cancer: a multi-level and a multi-causal phenomenon whose factors which have been identified over the last century. Scientists search for an explanation of cancer cells' latency, reversibility, multiple kinds of heterogeneity and metastatic properties. In this search, scientific practice tends to proceed by causal attributions, but the biological dynamism of the origin of cancer seems to be entangled with the very nature of living beings. The heterogeneity of the clinical manifestations of cancer and of the functional properties of tumour cells seems to prevent researchers from reaching the same study result, while some scholars foreshadow a unified theory of development, aging and cancer. Beyond the complexity of biological features and causal factors, some conceptual categories characterize the biology of cancer as a markedly organismic pathology. Spelling out such categories will help us understand the proposed explanations of cancer and the reasons why, although addressing the same scientific question about cancer origin, they seem to diverge.

In Chap. 2 I look at how the explanatory models of cancer have been changing to tackle the biology of cancer. The Somatic Mutation Theory (SMT), formulated in the 1980s, explains cancer as clonal expansion following genetic mutation. With the advancement of knowledge and experimental methods, the SMT weakened its strong assumptions, but survived through a wider category of models focused on the genetic and cellular levels, in which I recognize a Cell-Centred Perspective. The number of involved genes grew larger, as did the number of modes of genetic action; different molecular factors (e.g., epigenetic factors) entered the causal picture, and new concepts were coined (e.g., the Cancer Stem Cell, CSC). The importance of extra-genetic and extra-cellular factors became more and more apparent, and the causal role of interactions eventually prevailed, so that even the most important founders of the SMT ended up envisioning an incredible "wall of complication" with little clue of how this will be overcome.

Chapter 3 describes some systemic approaches to cancer. These approaches employ specific whole-genome or supra-cellular concepts and models to explain cancer origin. Of particular interest is the Dynamic Reciprocating Model (DRM) that works on the importance of the cell microhabitat (Extra-Cellular Matrix, ECM) and identifies crucial interactions that orient or disorient cell behaviour. Moreover, some models transform the genomic heterogeneity of cancer from anomaly (or noise) to causal factor. The concept of "field" (functional, morphogenetic) is shown to be a "trading zone" where different approaches to cancer converge, unless some different interpretations. In fact, systemic models in a sense complement the Cell-Centred Perspective: they work on the intricate web of discovered relationships, and they help in making sense of the properties of such network of interactions, leading to stability and instability, to "attractors" and to different kinds of dynamic equilibria. However, in this way systemic models also weaken the causal importance of specific mutations or molecules, or better, they provide a context of permissive *viability conditions* that may enable genetic and epigenetic causality.

For the first three chapters, the choice of categories of reference is a personal but not an arbitrary one. The bibliography is composed by the most cited authors, with the intention of remaining close to their interpretative vision. Their recognized authority further encouraged me to rely upon their work and discussion of empirical data.

In Chap. 4 I present the Tissue Organization Field Theory (TOFT), an anti-reductionist explanatory theory of carcinogenesis developed as a reaction to the assumptions of the SMT. The TOFT argues for a top-down causality in cancer, where the pathology is in tissue organization, and cell proliferation is a secondary derivation from abnormal signaling. I don't hide the hyperboles and contradictions of TOFT's polemic anti-reductionism. On the other hand, I value the epistemology proposed by TOFT, which will partly flow into my theoretical and epistemological proposals in the subsequent chapters. I accept the definition of the TOFT as a case of a wider Organism-Centred Perspective. I also offer some reasons for how the SMT and the TOFT might be compatible, arguing that the TOFT accommodates the cases in which the SMT works because the conditions for genetic determinism are created in the organic (cellular and tissue) microenvironment, so that it is possible to acknowledge *explanatory* independence and *epistemological* interdependence between the two theories.

In Chaps. 5 and 6, I formulate my positive proposal of a dynamic and relational view of cancer and outline the entailed features of biological explanation. I introduce several concepts which, I think, create the conditions for a more fruitful collaboration in the coming years of cancer research. I propose the new notion of Operational Integrating System. Unlike other system concepts, the Operational Integrating System does not take mereology – i.e., part-whole organization – as a baseline. This feature makes the notion more suitable to study the ways in which organization is maintained in space and time. Parts and wholes appear synchronically by a form of strong emergence, but the reciprocity between parts and wholes is also characterised by asymmetry, since the parts are always defined by means of the overall relationship and there is a causal dimension (named causality "by holding" as opposed to causality "by doing"), intrinsically relational, that stabilizes and guides the parts. I also coin the term "multi-unity dynamics," emphasizing the fact that both differentiation and state-holding processes are important, and that the imbalance between different kinds of causal dimensions might be at the root of the origin of cancer.

In light of the Dynamic and Relational View, biological explanation assumes some definite features. Explanation deals with biological determinations, i.e. regularities that hold dynamics. Biological determinations do not conflict with the deep indeterminism of biological entities; indeed, determination and indeterminism feed each other and are faces of the same coin. Explanation proceeds by finding "mesoscopic levels" (between macro and micro) where we can better understand the dynamics because we find the relationships that are more causally relevant to the *explanandum*. Reduction becomes reconceptualized as the identification of the *mesoscopic level*, and reductions in scientific practice are possible at determinate conditions. One of such conditions is the stability of the context that allows for "causal specificity," i.e., for the "causality by doing" by some particular part of the system (e.g., one genetic mutation). Anti-reductionist claims are here translated not in causal terms, but rather in epistemological terms, as dealing with the definition of the system and of the explanatory elements. In fact, my framework keeps two epistemological dimensions distinct: one is the definition of the system, the other is the definition of explanatory terms (or relata). Any (reductionist) explanation has an irreducible non-reductive dimension that concerns these defining endeavours. Parts are "essential-by-location" in the explanatory context.

In Chap. 7, I show that the same dual dimension is reflected (and recomposed) in the notion of "function" and in the hierarchical model of cancer, especially in the idea of a cancer stem cell. The "function" notion is crucial to any explanatory endeavor in the biology of cancer. In this context we find a discussion of evolutionary models of cancer, and the clarification of some problems that emerge in the most advanced reductionist models of cancer.

I have to add a final caveat. In previous versions of this volume, many pages were devoted to philosophical accounts of important contemporary philosophers who have developed influential accounts of causation, explanation, functions, reductions, emergence, etc. in the biological sciences. However, given the kind of methodology I have followed in this analysis, I have eventually decided to reduce that part to some notes and to a few pages which became the Appendix. Future works can, in fact, develop in much more detail a relevant critical analysis of different philosophical positions when confronted to the data and approach I have presented in this book. The seclusion of this background information, moreover, will facilitate a wider public in reading the book. Philosophical issues and different positions, when, nevertheless, cited and introduced, should be taken as part of the critical path I have followed, i.e., of a conceptual journey whose first map I hope to have just, but usefully, set up opened.

Acknowledgments and Sources

I wish to thank those who have been encouraging me to explore these issues in important exchanges, inviting me to visit other universities on various occasions. First of all, I thank Professor Marco Buzzoni, Alfredo Marcos, Silvia Caianiello, as philosophers, and Dr. Sylvie Menard, M. Luana Poeta and Alessandro Giuliani, as scientists, who have been following and orienting this study since the very beginning. Professors Juan José Sanguineti, Lucia Urbani Ulivi, Raffaella Campaner, Giovanni Boniolo in Italy, Maria Cerezo in Spain, and Sandra D. Mitchell, Jim Woodward, Jim Lennox, Ken Schaffner in Pittsburgh, Charles Wolfe and Jean Gayon in Paris, John Dupré in Exeter have also largely influenced my research and discussion of these topics. Their inspiring papers and literature have been of great help to outline the main philosophical questions that constitute the framework of this book. This work also profited from many discussions during my fellowship at the Center for Philosophy of Science, especially with Kyle Stanford, Maria Kronfeldner and Collin Rice. I am also grateful to Professors Carlos Sonnenschein and Ana Maria Soto for the opportunities they gave me to discuss and share with them papers and manuscripts, to Professor Robert A. Weinberg for the always prompt reply to any question and request of bibliography, and to the anonymous reviewers who have contributed to the progressive revision of the manuscript.

Finally, I wish to thank my family and friends whose support made this work possible. The discussions with them, also thanks to their different backgrounds – from engineering, to neuroscience, and from ethics to sociology – had great impact on my view of things and on the process of prioritizing the philosophical questions I have tried to address. A special thank has to go to Emanuele Serrelli who has done an incredible job in reviewing the final version of this volume. This work has been developed also as part of the research project funded by the Ministry of Science and Innovation of the Spanish Government; project title: "Change: Semantics and Metaphysics" (ref. FFI2009-13687-C02-01), and as a part of the project "Determinism and Indeterminism: From Science to Philosophy in Spanish-Speaking Academia" funded by the John Templeton Foundation (Grant #38920).

Chapter 1
Cancer Biology

1.1 Overview

Cancer is a multi-level and multi-causal phenomenon. It is also characterized by wild heterogeneity of effects, most strikingly affecting the same entities that take part in the neoplastic process. In this chapter I articulate these descriptive terms, presenting first the cross-level phenomenology of cancer (gene, genome, cell, cell-cell interaction, tissue, organism etc.). I devote particular attention to the distinction between two processes that, despite both being cellular processes, nonetheless encourage different levels of description: proliferation (how vigorously a cell makes copies of itself) and differentiation (what distribution of cell types emerges in a context over time). I then describe the importance of the second process, differentiation, as well as the parallelism between embryonic development and cancer, and the idea of a unified theory of development, aging and cancer. I then move to the multiplicity of causes that was always recognized for cancer, from the importance of the environment to the chemical theory, the parasitic theory and other theories, coming to the 'genetic turn' in the mid-1970s. I then treat tumour latency and tumour reversibility as possible evidence that cancer is characterized by 'causal complexity' more than by multiplicity of causes. I take causal complexity to mean not only that causes are many, but mainly that the causal factors are arranged in temporal dynamics which, in turn, influence them. Cancer complexity is in fact due to its being a dynamic process. Such complexity is revealed by the heterogeneity of cancer's clinical manifestations and of tumour cells' functional properties that prevents researchers from reaching a unified account of cancer. Moving to examine the multiplicity of effects of cancer, I then focus on the view of cancer as inter-level disregulation due to the uncoupling of processes (differentiation, apoptosis and proliferation). This is a particular view of cancer which emphasizes its dynamic aspect. I observe that dynamic views of cancer accommodate tumour heterogeneity, another typical effect of cancer. I spell out two different kinds of intra-tumour heterogeneity that coexist in the literature: H1, concerning the diversity of cells' differentiation stages

M. Bertolaso, *Philosophy of Cancer*, History, Philosophy and Theory of the Life Sciences 18, DOI 10.1007/978-94-024-0865-2_1

across the tumour, and H2, concerning genetic diversification. I draw some final implications before entering, in Chap. 2, into a review of how cancer research developed in pursuing the causal complexity introduced in this chapter.

Cancer research has been driven mainly by the question about the biological genesis of cancer in the organism: how do some cells come to proliferate and differentiate in a non-regulated and functionally integrated way? In the observed phenomenology, something goes wrong with respect to the normal behaviour of cells in a metazoan (any multicellular organism that develops from a zygote, i.e., a fertilized egg). Tumour cells, in fact, do not proliferate and differentiate normally, but give rise to abnormal tissues and masses whose cells may eventually invade other tissues and disrupt their functional structure as well. Why do cells in a tumour behave 'like this and not like that'?

Scientific practice usually pursues the question about tumour cells' behaviour by identifying factors – causes – that are involved in the origin, development and final onset of cancer. Such expectation dominates science and clinical practice. In the clinical tradition, the possibility of intervening in the main mechanisms of the disease is guaranteed by an adequate diagnosis (i.e., identification and distinction) and an understanding of the etiopathological and physiopathological aspects of the disease itself. In the same way, an inadequate understanding of the nature of the disease does not allow the identification of the most likely (or less likely) causes so as to predict the course of the disease, select the proper therapy for a specific patient, conduct clinical trials in at-risk populations and judge if the treatment prescribed is going to be effective. I wish to emphasize this point because, although the high heterogeneity of cancer's clinical manifestations and biological features could – and in fact often does – lead to renounce the notion of a unified definition and explanation of the disease, the success of major decisions in clinical practice depends on correct diagnostic and prognostic hypothesis. Due to the complexity of cancer, this ideal goal can be difficult and even impossible to achieve.

In this chapter I want to directly address: what features of cancer biology prevent us from grasping its original causes and getting, at least apparently, a common explanation? How should we understand cancer to make sense of such difficulties in getting a unified account of the different phenomenologies of this disease? Three main features demand attention and pose the main epistemological challenges: (i) the multi-level phenomenology of cancer, (ii) the multiplicity of causes involved in its origin and onset, (iii) the heterogeneity of tumour phenotypes.

1.2 Multiplicity of Levels

Usually described as an incurable disease, cancer was already known among ancient communities in Greece and Egypt as a multi-level phenomenon that affects tissues and organs, up to the whole functioning of the body. Since then, the aberrant morphological structure of tissues was the main diagnostic parameter for the pathology: the term 'cancer' was used to indicate an anomalous formation within some organ or tissue with a characteristic shape of a 'crab'. Such etymological derivation is preserved in a number of modern Latin-based languages as well as in German. The term 'tumour' (or 'neoplasm', from the Greek for 'new formation' or 'new growth') is often used as a synonym of cancer in common speech. The descriptive term is now confirmed through biopsy, i.e., the analysis of the histopathology of the compromised organ, and I will make the same use in this text, while not ignoring the distinctions that could be made.[1] As said before, my interest is to focus on the interpretation of the etiopathology of cancer to be intended as *a process* which is time dependent and, in some cases, reversible.

Multiple definitions of cancer became available over the last six decades. The lack of a unified causal definition of cancer reflects how different levels of the biological organization of a multicellular organism are eventually involved in the process of carcinogenesis, and the difficulty in tracking the number of different factors and their relative relevance. On the other hand, definitions of cancer are not merely related to the available technologies that explored the phenomenon of cancer at different levels – from genes, to cells and their interactions – but are determined by specific views of the intrinsic causal complexity of the neoplastic process.

Most definitions stress that cancer is an abnormal proliferation of a newly formed cellular mass, which may or may not be visible in an organism. This new mass, being no longer subject to the rules that control proliferation of the host tissue, invades it in a progressively disorganized fashion. Traditionally, cancer cells have been taken to continually divide and reproduce more than other cells of the same tissue, escaping the organism's control (IFO 2008). Cancer is "a heterogeneous disease often requiring a complexity of alterations to drive a normal cell to a malignancy and ultimately to a metastatic state" (Edelman et al. 2008). Indeed, some authors estimate that the generic term cancer can include more than 100 pathologies characterized by abnormal unregulated growth (Hanahan and Weinberg 2000). Cancer usually appears as "a disease involving dynamic changes in the genome; cancer cells have defects in regulatory circuits that govern normal cell proliferation and homeostasis (…) simplifying the nature of cancer, we can portray it as a cell-autonomous process intrinsic to the cancer cell, but cancer development depends upon changes in the heterotypic interactions between incipient tumor cells and their normal neighbours" (Hanahan and Weinberg 2000).

In the last 50 years, an emphasis emerged on cancer as "a disease of cell differentiation rather than multiplication" (Harris et al. 1969, 2004). An organicist vision in the last few decades has viewed cancer as "the result of the destruction of the tissue's architecture" (Sonnenschein and Soto 1999) or "a systems biology disease" (Hornberg et al. 2006), more than as a genetic or molecular pathology. Environmental and immunological factors have been integrated into the definition of cancer as well, reconstructing, to some extent, all the complexity of the original picture of the etiopathogenesis of cancer that had characterized scientific investigation at the

[1] The term tumour, in some literature, is used to refer both to a benign formation (i.e. a not (yet) invasive proliferating mass) and to an invasive cellular formation, in which cells show the typical invasive behavior that characterizes malignant tumours, or cancers.

beginning of the last century when both environmental and organismic factors were considered relevant in cancer origin and onset (for a more extensive review cf. Bertolaso 2011a, 2012a). Cancer is not considered as the cumulative result of genetic mutations producing defects in the circuits that regulate proliferation of normal cells. Interactions become more and more relevant and the dynamic dimension of the neoplastic process prevails over the presence or absence of certain molecular components (e.g., genetic mutations). This overriding role of interactions over parts leads sometimes to defining cancer as a systemic disease.

Finally, some definitions generalize from cancer and define it as "a non-adaptive process and a formless phenomenon" (Aranda 2002). The multi-level phenomenology of cancer has even led to focus on the disruption of functional properties of the organism, and to hypothesize intimate relationships between cancer and the ontogenetic process.

Historically, Van R. Potter was one of the authors who anticipated the idea of cancer as a problem intimately related to ontogenesis and to cell differentiation.[2] In Potter's time, the basic idea was that cancer cells have lost retroactive control mechanisms – or feedback control – of proliferation so they are able to divide unconditionally. Tumour cells then acquire a wide range of new properties that render them destructive of the organism as a whole. From studies of the biochemistry of cancer (Potter 1964), Potter imagined tumour growth as a problem of intercellular communication and differentiation, developing the concept of oncogenesis as "blocked ontogeny" (Potter 1968, 1969). He started by considering that phenotypic diversity in normal tissues may depend on two main domains: (1) evolution, where diversity emerges from adaptive DNA mutations in the genotype; (2) ontogeny, whose overall program of differentiation can be considered "as a process that alters the availability but not the information content of the total DNA complement" (Potter 1978). Enzymatic similarities in liver tumours and foetal liver tissue led Potter to suggest that phenotypic diversity in cancer could be explained without the need to assume important changes at the genomic level. Potter then proposed a vision of tumour cells based on the incorrect combination of the processes of proliferation and differentiation when the mechanisms blocking differentiation were not yet well known. Cancer was "a disease of differentiation", "a case of blocked ontogeny", where the block could conceivably occur at any stage between cell division and a state of terminal differentiation (Potter, cit.).

Subsequent studies have offered evidence that most, if not all, undifferentiated cells of mammals affected by myeloid leukaemia could be induced to differentiate into mature granulocytes and macrophages (Lotem and Sachs 1974), showing that the block was real, and in some situations could be overcome. The same was demonstrated for murine embryonic carcinoma cells whose differentiation could be induced by exposure to retinoic acid, analogues of cyclic AMP (cAMP), sodium

[2] Other authors could have been cited here, like Needham or Waddington. In fact, the idea that cancer was connected to development goes back to the nineteenth century. Potter is chosen for the particular relevance, for the current debate about cancer origin, of the vision he developed back in the 1970s.

butyrate, and other compounds, and for human acute promyelocytic cells (HL-60) obtained in culture, differentiated using a number of anticancer drugs, DMSO, vitamin D3, phorbol esters and analogues of retinoic acid (Ruddon 1995). Other studies have shown that blocking cell differentiation in a healthy organism induces a positive feedback, which increases the proliferation of cells upstream of the block, producing a hyperplasia. The block of differentiation induces the neoplastic transformation only indirectly. This was interpreted by some in terms of cell plasticity, as such blocks would be able to generate cells with different degrees of maturity and differentiation that could develop into tumours and constitute the neoplastic mass (Rapp et al. 2008). Successive mutations would be required to render their phenotype neoplastic. I will go back to the question of cells' plasticity in Chap. 5 (Sect. 5.3; see also 2.2).

1.2.1 A Unified Theory of Development, Ageing and Cancer

The cited data about similarities between carcinogenesis and ontogenesis paved the way for unified studies of their pathways and protein patterns. A simple example comes from studies of the WNT family proteins whose members – secreted glycoproteins modified by covalent bonds to lipids – are involved in embryogenesis, adult tissue homeostasis and carcinogenesis (Katoh 2008). Some authors even started to envisage a unified theory of development, ageing and cancer (Finkel et al. 2007; Soto et al. 2008a). The results of embryonic stem cell research have deepened our understanding of the mechanisms involved in the generation and assembly of tissues and organisms, including those related to ageing and tumourigenesis. Examples of regulatory mechanisms are micro-RNA dependent post-transcriptional regulation (Oakley and Van Zant 2007) and the epigenetic control of gene expression. Micro-RNAs (miRNAs) are small non-protein-coding RNAs that negatively regulate gene expression at the post-transcriptional level. Stem cells express specific profiles of miRNAs that, in turn, can alter the cells' differentiation potential. As for epigenetics, the stemness state of a cell appears to be correlated with its chromatin organization state and epigenetic modifications.

Ageing and cancer appear as deeply related as well. Some data on the role of stem cells in ageing suggest that stem cells age as a result of the alteration of processes that, over the course of life, work to prevent the onset of the neoplastic phenotype. Not only cellular factors that are inheritable through cell duplication (e.g., damage of the DNA), but also alterations in the niches that support stem cells, can contribute to the processes of ageing in mammals (Sharpless and De Pinho 2007).

Cancer has also been described as a chronic medical condition maintained by immunomodulatory factors (such as cytokines) that suppress immune function and ensure a microenvironment favourable to tumour formation by immunosuppression of the organism's control networks (Greten et al. 2004; Condeelis and Pollard 2006). On this basis tumours were also defined as "wounds that do not heal" (De Vita et al.

2008), raising awareness of the fact that the immune system ordinarily acts to prevent cancer.

Tumour cells display some characteristics that have been likened to embryonic cells. For example, they are less differentiated than their normal counterparts, and they divide rapidly and continuously and appear less specialized than fully differentiated cells. Some researchers studying leukaemia argue that the homeobox (HOX) genes, expressed during embryogenesis but down-regulated in adult life, are again expressed in some cancers. This hypothesis is summarized as follows: "oncology recapitulates ontogeny" (Grier et al. 2005). The fusion of HOX genes caused by chromosomal translocation compromises[3] the regulation of pluripotent stem cells, modifying their normal process of differentiation and forming tumour-producing progenitors. Although our understanding of the molecular and cellular basis of the development of the prostate is still too limited to definitively validate this hypothesis, there are some data common to other types of tumours, which show that many of the genes that regulate the embryonic development of this organ are again expressed during neoplastic progression of prostate tumours (Marker 2008). Evidence that, in the early stages of testicular tumours, the precursor of the neoplasm develops from a gonocyte with stem capacity would explain how, in an adult man, structures that appear as "neoplastic caricatures of embryonic growth" can develop (Skakkebék et al. 1998).

If tumourigenesis and embryogenesis are similar under certain respects, important differences must also be acknowledged, as demonstrated by experiments on the differential effects of the same mutation during embryonic differentiation and neoplastic transformation (Biava 1999). Such context-dependence of the effects of genetic mutations leads to a consideration that will be fundamental in this book: the pathologic character of tumour cells goes beyond any genetic or biochemical alteration (Biava 2002). Tumour cells have defects in regulatory mechanisms of differentiation, being unable to read correctly the signals they receive. The onset of the neoplastic phenotype seems to be related to the inability of stem cells to differentiate in certain conditions of the microenvironment.

The abovementioned development of tumours resulting from disturbances in the process of cell differentiation, caused, in turn, by chromosomal translocation, has been defined as cells acquiring a function. But the most noticeable difference between normal and tumour tissue lies in the imbalance between the processes of cell differentiation and proliferation, allowing tumours to produce an accumulation of aberrant undifferentiated, or partially differentiated, mitotically active cells. During embryogenesis there is, in fact, a fine balance between cell proliferation and differentiation essential for the normal development of the foetus, whereas in cancer it is precisely the balance between the two processes that is compromised as it is not brought to a successful completion (Abbs et al. 2004). These findings reinforce evidence of the multi-level phenomenology of cancer and its causal complexity while emphasizing the regulatory dimension of the overall process. Recent research on the

[3] However, the functional relationship between leukaemia and deregulated activation of HOX genes is still unclear.

early development of prostate cancer supports this idea (Marker 2008). The different rate of proliferation and degree of differentiation of cells that give rise to the phenotypic differences and the metabolic changes that are found in tumours are linked to the heterogeneity that characterizes the cells of a tumour mass.

The consideration of the fascinating possibility of a unified theory of development, ageing and cancer leads to an observation by Rubin: "[c]ancer is an enormously complex biological phenomena that needs to be considered at multiple levels to achieve reasonable understanding" (Rubin 1999). The multi-level phenomenology of cancer surely involves cell differentiation and its regulation in the organism, as we will see in more detail in the next Sections.

1.3 Multiplicity of Causes

Cancer was originally considered as a disease related to the environment as much as to endogenous factors. This pathology was not the object of more specific studies until the end of the Nineteenth Century and the beginning of the Twentieth Century, when its direct relationship with the genetic and cellular component was found. Rudolf Virchow (1821–1902), who in 1863 published a treatise in which tumours were classified according to their morphology, was the first to purport the idea that diseases, especially cancer, are at the same time both natural and social events, which are generated on the one hand by "incorrect" nature and on the other hand by the "excesses of the environment". At that time importance was assigned to the discovery made a century earlier by the English surgeon, Percival Pott (1714–1788). He claimed that cancer of the scrotum, which frequently affects chimney sweeps, was due to soot residues deposited in that area of the body. Pott was not the first scholar to establish a link between environment and disease. This factor had been discussed previously. In his work, Bernardino Ramazzini (1633–1714) described clinical aspects of patients with "work related illness" and those who suffer health damage "by virtue of unhealthy lifestyle" as, for example, the use of tobacco powder. The medical scientific research of the late 1800s, therefore, not only emphasized, following Claude Bernard, how endogenous causes (*milieu intérieur*) can lead to disease, but also sought to identify external factors arising from socio-economic conditions (*milieu extérieur*). Beyond this distinction that would require a much more detailed and careful historical analysis, I wish to point out that with Virchow social medicine was born.

Later on, the development of new biochemical tools, with the aid of the microscope, and the possibilities to enlarge analysed data emerging from the environment led to finer-grained studies of cancer's pathological features. Several hypotheses on the origin of cancer began to emerge ranging from the chemical theory, which raises the possibility of an alteration of the cell's biological balance caused by toxic substances in the environment, to the popular parasitic theory and to the cellular evolution theory. In the Nineteen-Fifties and Sixties, the formulation of the genetic theory of cancer took over the work of the scientific community – microbiologists, pathologists, biologists, clinicians and surgeons – while the relevance of the organic environment and of the immune system was progressively set-aside for a time. New observations seemed to confirm the idea that cancer could be a genetic-cellular pathology. Microscopic technologies revealed a high disorganization of chromatin in cancer cells, and this added a new level of structural and morphological observation of the disorganization of the tumour masses, a level that – interestingly – was later confirmed by discoveries on DNA and the molecular basis of genetic inheritance.

The studies by Boveri on sea urchins' eggs contributed to this vision. He devoted considerable energy to study the association between aberrant mitosis and cancer (Boveri 1914) using experimental manipulations of this type of eggs and inducing multipolar mitosis and aberrant chromosome segregation. The unlimited growth that resulted – and that is commonly associated with the malignant tumour phenotype – was attributed to aberrant chromosomal arrangement. For some this approach laid the foundation for the view of cancer as a genetic disorder.[4] However, it was only in 1960 that the first genetic defect associated with cancer was identified, illustrated more in detail in the next Chapter (in particular Sect. 2.2): it was visualized as a small chromosome in cells of patients with chronic myeloid leukaemia (Nowell and Hungerford 1960). Subsequently, in 1980, David von Hansemann published the first mitotic figures from 13 different tumour samples, all characterized by aberrant chromosome structures.

This evolution of cancer research – labeled the "genetic turn" – was reinforced by technological progress and by the discoveries of the double helix of DNA and of the first mechanisms of DNA duplication and transmission. Moreover, the discovery of new correlations between defects in DNA repair and a higher predisposition to cancer reinforced the hypothesis that cancer was a genetic disease and *in vitro* experiments – in which mammalian cells deficient in DNA repair mechanisms had an increased susceptibility to malignant transformation by physical or chemical carcinogens – seemed to support this genetic turn as well (cf. Cleaver 2005; Wijnhoven et al. 2007). Therefore, cancer research focused on genes that had the potential to cause cancer. Ras and src genes (Duesberg 1980) were among the first to be identified by cloning technologies (Tabin et al. 1982). These data, all together, found broad approval and contributed significantly to an interpretation of the molecular basis of cancer.

Although molecular and genetic studies took over the challenge to explain cancer and to understand its mechanisms while, little by little, the interest for the initial

[4] Some authors claim that Boveri might be considered a reductionist (Soto and Sonnenschein 2004) in the sense clarified in the Appendix. I do not think that this is necessarily the case. Mainly, he tried to explain a cellular behavior in terms of the disruption of the genetic patrimony of the cell. He did not claim that the DNA as such might be able to account for the whole process and cellular transformation. From an epistemological point of view, the two claims are different: the first one is a claim about how to study biological (dis)functions; the latter is a claim about genetic determinism. Gayon (2006) discusses this difference in a very convincing way. Further discussion require a deeper understanding of what a reduction in science is (see Chap. 6) and how biological determinations should be understood.

pathological studies on the nature of neoplastic mass dynamics and its cellular composition waned, the aetiology of cancer remained at the forefront of epidemiological research programmes, and the relevance of both environmental and biological factors in carcinogenesis was never completely overlooked in this field. Several historical studies showed that environmental factors, lifestyle and genetics might be players in tumour onset and its metastatic progression. A typical example is that of impact studies of cancer in female survivors of the atomic bombs of World War II (Tokunaga et al. 1979). Tumours arose only after a period of time and in an almost synchronous manner for many of those who had been exposed to atomic bomb radiation. The multiplicity of causes of cancer was strongly shadowed by the genetic turn and by the consequent emphasis on genetic mutations and molecular biology. But, as we have just seen, the multiplicity of causes never completely succumbed to this narrowing of cancer research as a whole.

The disruption at the level of nucleotide sequence was considered the main cause of neoplastic onset and progression (Luch 2006), and *in vitro* experiments showed that DNA was the common denominator between different kinds of carcinogenesis (e.g., physical and chemical). Nevertheless, subsequent studies highlighted that tissue injury too can induce the formation of a tumour in cells that are located at the edges of the wound. Even something as trivial as injuring skin that had been exposed to an initiator carcinogen can spark cancer development, suggesting that events related to tissue organization could be sufficient to achieve the same effect. As we will see more extensively in the next chapters, this hypothesis is supported by more recent evidence that no mutagenic initiator or somatic mutation is necessary to give rise to a neoplastic phenotype (Hendrix et al. 2007), but it is likely to be the cancer itself, in some way, that induces mutations; "it may be more correct to say that cancers beget mutations than it is to say that mutations beget cancers" (Prehn 1994, p. 5296). In this way, important questions about the regulatory pathways involved in cancer origin and development arose, independently of the molecular parts involved.

1.3.1 Latency and Tumour Reversibility

As I have said in Sect. 1.1, multiplicity of causes means that there are many causal factors responsible for cancer on which researchers can focus. Causal complexity means, instead, that different kinds of causal dependencies coexist, and that the causal factors are themselves influenced by the temporal dynamics in which they are arranged.

The course of cancer is not predictable with a high degree of certainty and varies greatly between different patients, even with the same tumour type. The variables that are relevant in the timing of the neoplastic process vary a lot depending on the single cases. Consistently, the appearance of the clinical symptoms of a tumour coincides not with cancer origin but rather with the terminal phase of the neoplastic disease, preceded by a variable period of time called the *latency period*. Throughout the period of latency, which in many human tumours can last years, the tumour

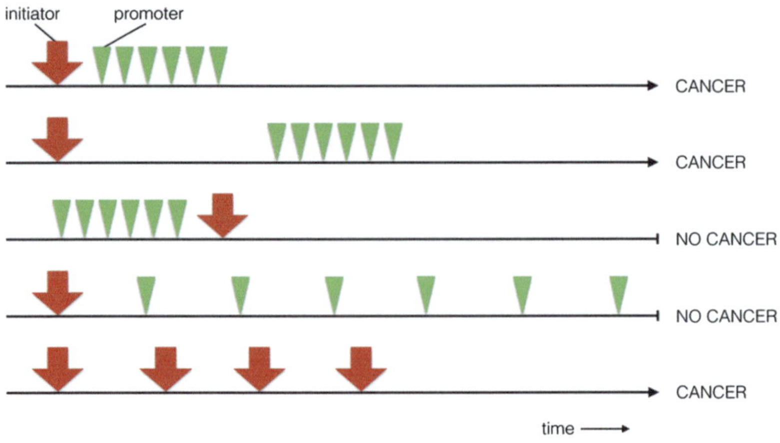

Fig. 1.1 Graphical representation of the importance of sequence and frequency of initiator and promoter events for tumours' initiation and progression (Modified from Alberts et al. 2002)

already exists as an aggregation of neoplastic cells in the process of replicating, but is not yet clinically identifiable. Also for this reason, prevention and early cancer diagnosis are of great interest in the clinical practice, so that cancers can be detected in their preclinical stage, that is, during the latency phase.

Timing is central in the neoplastic phenomenon. Some of the most compelling evidence for this dates back to studies conducted in the 1970s, on the carcinogenic effects of chemicals applied to the skin of mice. These animals develop skin cancer if repeatedly exposed to potentially mutagenic chemical carcinogens, such as benzopyrene or its analogue, dimethylbenzoanthracene. It quickly became evident, however, that a single application of the carcinogen was not, by itself, sufficient to give rise to a tumour or to any other anomaly. The predominant hypothesis was that treatments generate a latent form of genetic damage, a mutation that paves the way for increased susceptibility to cancer, if the cells are exposed to the same substance or other compounds (albeit very different in terms of aggressiveness). Alternatively, exposure for months to substances that work as promoters but are non-mutagenic can also cause skin cancer in areas previously exposed to tumour initiators. An efficient representation of these mechanisms of initiation and promotion was first reported by Boutwell (1978), who presented the result, in terms of neoplastic onset, of successive exposures to a mutagenic tumour initiator, and to a non-mutagenic tumour promoter. A key finding was that both the *sequence* and the *frequency* of these exposures were relevant to the onset of cancer (Fig. 1.1). *In vivo* experiments on tumour formation had shown that carcinogenesis was a process led by a sequence of events in a specific order integrated at cellular levels through molecular signals.

Phenomena of tumour regression likewise demanded for a reflection on the tumour phenotype's dependency upon temporal and micro-environmental factors. A clarification is urgent, given the possibility of a tumour reversion is strictly linked to the features of the regulatory dynamics at work in carcinogenesis and commonly acknowledged as one of the most relevant issues at stake in making sense of tumour cells behaviour and nature (NCI 1976; Challis and Stam 1990; Baker and Kramer 2007). Well known experimental data show that tumour cells can be normalized if introduced in a normal microenvironment, like a blastocyst (Mintz and Illmensee 1975; Hochedlinger et al. 2004), or like in the case of cells from hepatocarcinomas injected into a normal liver, or if their extracellular components are modified (McCullough et al. 1998; Bissell and Radisky 2001).

1.4 Multiplicity of Effects

We have seen that from an epidemiological point of view cancer is considered a complex process because it involves several stages where numerous events partake in each step. The disease exhibits remarkable heterogeneity of biological and clinical features linked to its genetic and environmental complex aetiology, so that, as De Vita has highlighted, you have to consider "geneN-environmentN interactions, which for how many 'n' is not known" (De Vita et al. 2008). For this reason, characterizing each specific risk factor, within a broad framework of factors involved becomes difficult and also limits greatly the possibilities for statistical analysis. Cancer occurs at any age and, as we mentioned in 1.3, in the early stages of proliferative and cellular accumulation, it is generally asymptomatic.

To further complicate the matter, there is a wide multiplicity and heterogeneity of cancer effects, starting from the fact that almost all of the symptoms caused by cancer can be commonly associated also with non-neoplastic diseases or various types of internal pain (Holland 2003). Even the wide variability of individual response to carcinogenic substance exposure indicates that the reactions are not homogeneous, reason for which the experimental and epidemiological models cannot be considered as representative of all subjects of a given population. Finally, therapeutic responses show a high inter-individual variability and depend on many things, including psychological, behavioural and social factors (Yale et al. 2005).

The most surprising heterogeneity in cancer is, however, found inside the tumour itself. The cells appear different from each other. Before the genetic era of cancer, evidence for heterogeneity had been described in terms of cell morphology, tumour histology, karyotypic and cytogenetic markers, rates of cell growth, cellular by-products, reactors, enzymes, immunological characteristics, metastatic capacity and sensitivity to therapeutic agents for different neoplastic phenomena (Foulds 1954; Heppner 1984; Heppner and Miller 1998; Dexter et al. 1978). Observations that tumours seem to contain a heterogeneous population of cells were explained by postulating changes in the tumour microenvironment and coexistence, therefore originating from different sub-clones genetically created by the progressive accumulation of independent somatic mutations. An additional explanation however – coherent with some considerations on the similarities in cancer and organogenic processes – hypothesizes that human tumours are not simple clonal expansions of transformed cells, but rather a kind of complex three-dimensional tissue where the cancer cells can become functionally heterogeneous as a result of a spatial-temporal aberrant differentiation process. According to this scenario, where temporal and micro-environmental factors play an important role, "tumors act as caricatures of their corresponding normal tissues and are sustained in their growth by a pathological counterpart of normal adult stem cells, cancer stem cells" (Dalerba et al. 2007, p. 267). Recall the image mentioned in Sect. 1.2.1, of cancer as a caricature of embryonic growth at tissue level. This model, initially developed for human myeloid leukaemia, in recent years has been extended to other types of cancers, such as breast or brain cancer (cf. Sect. 2.6).

1.4.1 Intra-level (Dis)Regulation as (Un)Coupled Dynamics

As mentioned above, by the 1980s research papers started to put together and integrate key new molecular elements to explain cancer, by characterizing the carcinogenic process from the point of view of regulatory processes. The loss of certain cellular properties began to be seen not so much in correlation with the diversity of molecular causes of tumour formation, but also with the *alteration* or *decoupling of some mechanisms*. Differentiation, apoptosis and proliferation constitute the main elements of this explanatory axis. This change of emphasis had a relevant influence in scientific practice and allowed a better understanding of tumour cell heterogeneity.

The main characteristic of tumours remains the recklessness of cell proliferation, generating an accumulation of cells with aberrant phenotypes within the tumour mass. At the same time, the disruption of the organisational features of tissues in cancer is due to a deregulated balance between the apoptotic process and the process of progressive cell differentiation within the tissue. Receptors for apoptosis are ordinarily in charge of reporting to the cell environmental situations that are not compatible with life, resulting in the induction of programmed cell death. Although the biochemistry of apoptosis is well studied, including its mechanisms (Ashkenazi and Dixit 1998) and molecular factors (Green 1999, Vaux and Korsmeyer 1999), it is not always clear how and why the apoptotic process is circumvented by tumour cells. Nevertheless, some cases have been well described, such as the effects of the loss of the adhesion between the cells and the Extra-Cellular Matrix (ECM, see more in Sects. 2.8, 3.3.3, 4.2.2, 4.3, 5.3.1). Researchers showed that this adhesion strongly influences the normal functions of growth, differentiation and proliferation of the cells. This dependence is usually indicated by the term "dependence from anchorage". Normal cells that separate from their ECM anchoring system usually promptly undergo apoptosis,[5] whereas tumour cells proliferate and spread.

[5]This form of programmed cell death is known as "*anoikis*" and is induced when cells detach from the ECM (Frisch and Screaton 2001). Usually the cells remain connected to other tissue components, allowing communication between adjacent cells and between them and the ECM, which

These findings and studies reinforced the idea that in the onset and stabilization of the neoplastic phenotype new balances between cell proliferation and differentiation are at stake, altering the normal integrated functioning of biological structures like cells and tissues, as already mentioned in Sect. 1.2.1. The loss of some cellular properties and the neoplastic phenotype can thus be studied in terms of the process of cell proliferation and of uncoupled cell differentiation or apoptosis rather than in terms of alterations of molecular factors. In evolutionary terms, in fact, the apoptotic process is actually part of the differentiation process. Consequently apoptotic processes remain present, albeit in aberrant form, in various stages of neoplastic transformation. The point is that in actively proliferating tumour cells it is *as if* the genes, whose protein products are required for regulating cell proliferation, were stuck in the "on" position, when they should be able to switch to the "off" position in a normal process of cell differentiation. This phenomenon is due to the fact that the genes linked to differentiation control are either not expressed or are expressed in an abnormal manner.

1.4.2 Tumour Cell Heterogeneity

Earlier I pointed to intra-tumour heterogeneity as the most striking heterogeneity in cancer biology. In the literature, two different kinds of heterogeneity have been extensively referred to. One is related to the aberrant differentiation of cells within a tumour (H1), the other to the genetic drift of tumour cells that end up with very different genetic configurations, phenotypes and functional states (H2). To better understand how H1 and H2 are related and how they have been described, let us go again through some historical steps of cancer research that we have already touched on above.

Traditionally, cancer cells have been portrayed as reasonably homogeneous cell populations until when, in the course of tumour progression, hyper-proliferation, combined with increased genetic instability, spawn distinct clonal subpopulations. Reflecting such clonal (genetic) heterogeneity, many human tumours are histopathologically diverse, containing areas at various degrees of differentiation, proliferation, vascularity, inflammation and invasiveness. In recent years, the panorama has been further clarified. "(E)vidence has accumulated pointing to the existence of a new dimension of intra tumour heterogeneity and a hitherto-unappreciated subclass of neoplastic cells within tumours, termed cancer stem cells" (Hanahan and Weinberg 2011, p. 662). The scheme of stem cells – partially differentiated transit-amplifying cells, and fully differentiated end-stage cells – seems to be recapitulated in many carcinomas and other tumour types (Ailles and Weissman 2007). The stem-like characterization of these cells suggests that a single, genetically homogeneous population of cells within a tumour may nevertheless be phenotypically

supplies fundamental signals for survival and growth. When the cells are detached from ECM therefore, for example, because of a loss of the normal interactions, they undergo *anoikis*.

heterogeneous due to the presence of cells in distinct and non-coordinated states of differentiation. Just like a normal stem cell a progenitor cell might be able to reproduce and also produce cells that differentiate aberrantly. Phenotypic changes are usually mediated in this process of differentiation by non-genetic events. Epigenetic factors are often at work. By H1 I refer to the heterogeneity that characterizes differences among cells at different aberrant stages of differentiation.

H2, instead, refers to the equally important source of cellular variability in cancer derived from genetic heterogeneity in a tumour that accumulates as cancer progresses. Different mechanisms can be invoked to account for H2. Elevated "genetic instability operating in later stages of tumour progression is correlated with rampant genetic diversification that outpaces the process of Darwinian selection, generating genetically distinct subpopulations far more rapidly than they can be eliminated" (Hanahan and Weinberg 2011). Such thinking is increasingly supported by in-depth sequence analysis of tumour cell genomes, which has revealed striking intra-tumoural genetic heterogeneity through the sequencing of the genomes of micro dissected cancer cells taken from different sectors of the same tumour (Yachida et al. 2010).

For the moment I leave aside discussing to what extent Darwinian selection can really account for the neoplastic process. I will get back to that in Sects. 2.8 and 7.2.3 (for a quick overview see Huang et al. 2009 and Bertolaso 2009b). Here I am interested in stressing that tumours heterogeneity makes it apparent how different levels of organization are lost in the neoplastic process. Given both H1 (related to abnormal cells differentiation: the tissue organizational level) and H2 (related to tumour cells' instability: the genomic level), the processes of increasing tumour heterogeneity imply the disruption of the correct processes of differentiation in the progenitor tumour cell (see Sect. 2.5) with a concomitant loss of control on the genomic level of organization in tumour cells. A graphic description is presented in Fig. 1.2. The non-linearity that characterizes the mathematical models that try to formalize the complexity of interactions (for a review and examples cf. Bizzarri et al. 2008; Vineis et al. 2010; Cherubini et al. 2011) justifies the difficulties in calculating the probability of a tumour arising and manifesting itself. However, beyond the non-linearity of causal combinations and interactions, cancer complexity opens the question on the more general dynamism that governs the organization of the multicellular organism, with its morphostatic and morphogenetic dimensions. Molecular events and biological processes seem intertwined in generating and guiding the progression of this phenomenon, according to a space-time regulatory dimension.

1.5 Conclusion

The dynamic features of cancer are reflected in the different kinds of tumour heterogeneity and in particular in the intra-tumour heterogeneity. Central to these features is the process of cell differentiation and development. In metazoans (multicellular

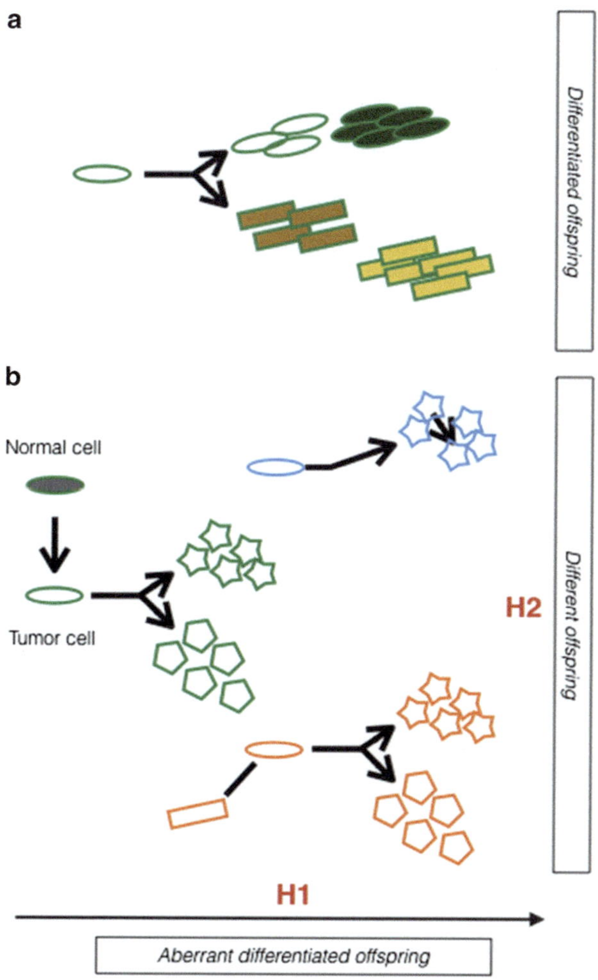

Fig. 1.2 (**a**) Differentiation can be considered the sum of the processes by which cells of a multicellular organism reach their specific function. Through the acquisition of specific functional features, the offspring can be distinguished from progenitor cells and from other cells in the body belonging to different tissues; (**b**) a tumour cell starts proliferating in an abnormal way, giving rise to cell heterogeneity that simulates an aberrant tissue differentiation (H1). However, as far as the neoplastic process proceeds, new kinds (different colours) of tumour cells arise from a general instability of the cellular genome (H2) (Redrawn from Bertolaso 2012a)

organisms), the process of differentiation appears as a virtually permanent development: some cells actually retain the ability to divide, while others divide and differentiate into cells with much more restricted phenotypic plasticity; these are called pluripotent instead of totipotent cells as they are already committed towards specific tissue dependent function. The multilevel phenomenology of cancer implies that we are not faced with the mere inexorable progression towards a final epilogue through discrete events but with a process involving different scales and a continuum of events, where the passage from one to another is determined by numerous internal and external organismic factors.

As we will see in Chap. 2, explanatory models have evolved over time, trying to make sense of the complexity of cancer and of how the parts-whole (cells-tissue) relationship is reshaped in this process. To develop adequate tools to control the disease, scientific research attempted to get a causal account of it. After all, in common scientific practice, providing an explanation coincides, to some extent, with providing information about causes (cf. Woodward 2010). But multiplicity reigns in cancer: not only of cancer definitions, but of causal accounts. This multiplicity will require the consideration of the non-obvious relationship between explanatory accounts and causal attributions in scientific practice.

Chapter 2
The Evolution of Explanatory Models of Cancer

2.1 Overview

Robert A. Weinberg is undoubtedly an important figure in cancer research. He is best known for discovering one of the first human oncogenes, Ras, and the first tumour suppressor gene, Rb, and for his 2000 seminal paper with Douglas Hanahan "The Hallmarks of Cancer". In 2014 Weinberg wrote a "Leading Edge Essay" celebrating 40 years of the journal *Cell*. We will turn to that article below. For now, let us consider Weinberg's impression about the last 40 years of cancer research:

> Those who have participated in cancer research during this period have witnessed wild fluctuations from times where endless inexplicable phenomenology reigned supreme to periods of reductionist triumphalism and, in recent years, to a move back to confronting the endless complexity of this disease (Weinberg 2014, p. 267).

In his paper, Weinberg simplifies the history of cancer research in "phases" or "periods". Cancer research appears as a field historically very sensitive to paradigmatic cases, i.e., discoveries that open new lines of research that then become mainstream and are followed by many laboratories and financially supported. This pattern in the evolution of cancer research makes it possible to find "key" general models; models that were important by opening new lines of research, and gross periodizations. Furthermore, often key models are characterized by an attention to one or a few entities or processes, potential targets of cancer diagnosis and therapy. In the 1970s cancer research in the United States was bound to the conviction that "the key to understanding human cancer" was the reverse transcriptase enzyme (Weinberg 2014, pp. 266–267). Indeed, this conviction characterized the U.S. War on Cancer that, thanks to Richard Nixon's 1971 National Cancer Act, made "an enormous pot of money" suddenly available for cancer research. In agreement with this periodizing approach (and with the idea of "key models"), in this chapter I look at how the explanatory models of carcinogenesis have been changing to tackle the features of the biology of cancer presented in Chap. 1. Another feature of cancer research that I exploit in this chapter is that several researchers have explicitly

M. Bertolaso, *Philosophy of Cancer*, History, Philosophy and Theory of the Life Sciences 18, DOI 10.1007/978-94-024-0865-2_2

categorized their models and addressed their epistemological assumptions. This chapter favours such categorization, without however ignoring important contributions to scientific research by other authors.

The interest of this chapter is both historical and epistemological. I analyse the scientific literature to see which elements have been integrated into the explanatory models and why. Should the Kuhnian notion of "paradigm" be applicable, we might recognize a dominant Cell-Centred Perspective all the way through the models, while they go by considering more and more diverse causal factors. Over years 1980–2000, as Weinberg has admitted, "a flurry of molecular and genetic research gave hope that cancer really could be understood through simple and logical reductionist thinking", then molecular biology itself led to complicate the picture so much that now "[w]e lack the conceptual paradigms and computational strategies for dealing with this complexity" and "we can't really assimilate and interpret most of the data that we accumulate" (p. 271). So, to understand the future of cancer research, we should first recognize that the gene-cell centred viewpoint, after decades of productive research, is now bumping against a wall of complication.

2.2 The Somatic Mutation Theory and Its Origin

As we have seen briefly in Sect. 1.3, at the beginning of the Twentieth century the study of the etiopathogenesis of cancer, driven by epidemiological findings, was concentrated on physical and chemical causes (also reviewed in Parkin 2004; Colditz et al. 2006). In those years, Boveri's studies showed a correlation between aberrant mitosis and malignant tumour formation, suggesting the idea that the causes of cancer resided at the molecular genetic level. Nevertheless, the association of the first clearly genetic defect with cancer had to wait until the late 1950s, when a small chromosome was identified in cells from patients with chronic myeloid leukaemia (Nowell and Hungerford 1960). In the same years, the discovery of the DNA as the molecular basis of heredity led to a re-evaluation of the pioneering findings of Peyton Rous on the ability of a tumour of birds to generate another tumour when transplanted into a healthy individual, thus showing that a biological agent – as it was called – could directly cause tumours (Rous 1910). These classic findings now suggested that the tumour-causing virus contained a gene that conferred tumourigenic properties to cells. Viruses were now identified by their ability to induce tumours when inoculated in a host (reviewed in Klein 2002). By the mid-1970, as Weinberg puts it, "[h]alf a century of cancer research had generated an enormous body of observations about the behavior of the disease, but there were essentially no insights into how the disease begins and progresses to its life-threatening conclusions" (Weinberg 2014, p. 267).

The molecular biology of cancer began in the 1970s, although at the beginning, "the notion that cancer was a disease of identifiable genes was little more than an attractive speculation" (Weinberg 2014, p. 268). Then the Varmus-Bishop discovery of the *src* proto-oncogene in 1975–1976 and other studies had a fundamental impact on cancer research, connecting genetic mutation to cell transformation, and directing work towards identifying the genetic components of cancer progression. Subsequent studies revealed how genes, carried by these tumours and referred to as Oncogenes (ONG), might be considered responsible for cancer by encoding proteins involved in cell growth regulation and proliferation or in signals transduction. The idea that mutations in the cellular homologues of viral genes could transform cells without the involvement of a virus matured and grounded the idea that a molecular (mutated) gene could be considered tumourigenic in its own right. Findings of particular importance demonstrated that the neoplastic phenotype could be transferred when the DNA from a mutated and transformed mouse cell was transferred into non-neoplastic cells. At the same time, some researchers began to study the relationship between oncogenes and the genes encoding proteins involved in cellular differentiation and development (Shih et al. 1979; Cooper et al. 1980). In 1982 the Weinberg group cloned the first ONG from a bladder carcinoma cell line, after having isolated the DNA fragment considered responsible for cancer through a series of transfections.[1] These cloned cellular genes showed the same properties of transforming ONG from retroviruses. The 1982 discoveries were crucial to route modern cancer research: the complex interplay of different types of genetic lesions (Shih and Weinberg 1982) and the role of ONGs in cell transformation became the main focus of many research programmes (Tabin et al. 1982). The concept of *mutation*, which in its most general sense means simply a change in the DNA sequence, was introduced at that time to explain the change in the functional identity of somatic cells from normal to cancerous.

The explanation of cancer as a problem of clonal expansion of tumour cells mediated by somatic mutations, i.e., alterations of the DNA sequence of the cells that constitute a multicellular organism, emerged in those years. That explanatory approach is now known as the *Somatic Mutation Theory* (SMT), and still stands as a major reference point. The term "somatic cell" is used in contrast to "germ cells", i.e. the cells that are involved in the reproduction of multicellular organisms. The clonality of tumour cells and the determinant role of somatic mutations are the fundamental tenets of the SMT. The biological assumptions can be consistently listed as follows: (1) after initial differentiation of the zygote, the metazoan cell settles down on a default state of quiescence; (2) mutations in genes involved in the neoplastic development increase the rate of proliferation of neoplastic cells; (3) carcinogenesis thus originates at the level of the single cell; (4) neoplasms fundamentally

[1] The scientific payoff of transfection methodologies employed in the 1980s has been largely criticized. Further details and specifications are beyond the aim of the present volume (for a review see Blanpain 2013). I think that a more nuanced analysis is needed to understand why and to what extent those procedures are useful to draw conclusions in the scientific field. In particular, my suggestion would be to consider the relevance of control in experiments and the epistemological status of experiments.

come from single clones; (5) neoplasms arise when the genes involved in controlling proliferation have undergone mutation.[2]

By the time the SMT was formulated, available evidence indicated that an oncogenic transformation of cells in a primary tumour involved at least two stages: initiation – identified by the immortalization[3] of cells – and cell transformation.[4] A whole long-term research programme was launched to search for the key mechanisms of neoplastic transformation. Indeed, the discretization of cancer went further by leading, at least in educational settings, to the idea that the neoplastic process can be understood in term of *discrete events*, more specifically: initiation, promotion, transformation, progression (according to the nomenclature originally elaborated by Rous).[5]

The term 'initiation' refers to the first alteration that begins to deregulate a series of pathways. Still far from inducing a real neoplastic transformation, the altered pathways may be manifested in normal or abnormal proliferation, and in the orientation towards differentiation and death, also depending on the individual's genetic characteristics. The initiation of a tumour cell must therefore, in some way, alter some mechanisms through genetic mutations that can be familial, i.e., the individual is born with an initiator alteration in his or her genome. According to this model, any type of carcinogen should cause, at some point, a mutation compromising the pathways that usually trigger the activation of the first DNA repair mechanisms designated to the maintenance of correct signal transductions. In this way, both, self-destruction mechanisms by activating cellular apoptosis, and DNA repair, can be compromised and fail so that the first mutations arise and can be transmitted to the cells' offspring.

The 'promotion' phase takes place when, through exogenous stimuli, cell proliferation initiates and mutations, due to replication errors, confer on the cell a 'selective advantage'. The cell begins to accumulate a series of genetic errors involving, for example, ONGs, tumour suppressor genes and other genes that control apoptosis. Immunoediting (the recognition of these cells as "non-self" and their destruction) plays an important role in this second stage, and in some cases stops tumour progression. When a tumour cell emerges from this phase, it shows dis-regulated

[2] These features were pointed out by opponents of the SMT (Sonnenschein and Soto 1999). As we will see also in Chapter 4, supporters of SMT do not deny this schematization. For them, although emphases may vary, cancer mainly remains a sub-cellular molecular and genetic problem.

[3] By "immortalization", I mean the production of a cell line capable of an unlimited number of cell divisions. Immortalization can be the result of a chemical or viral transformation or of fusion of the original cells with cells of a tumour line.

[4] Cell transformation is the change that a normal cell undergoes as it becomes malignant.

[5] This attention to genetic mutations and molecular mechanisms always struggled with the evidence that tumour cells change continuously during the neoplastic process, making it difficult to predict their course by using only genotypic or cytological analyses of the tumour cell, to make prognoses regarding how the pathology will progress, or even to describe them. The progressive changes seen in tumour cells motivated an early eccentric line of questioning upon which level of biological complexity is most adequate to study the phenomenon. I will turn to these issues again in the next Chapters. Some questions have been presented already in Sects. 1.2 and 1.4.2.

proliferation or differentiation pathways and often alterations in the normal process of cell death. At this point the angiogenesis and evasion from immune surveillance begins, thus allowing the cell to enter the stage of neoplastic transformation.

The neoplastic 'transformation' is linked to the ability of tumour cells to grow without being subject to, or requiring, any external control. The tumour of monoclonal origin consists of cells with characteristic high phenotypic plasticity, understood as the ability to continue to change their physical aspect and structure in response to the external environment. Other mutations come into the picture at this point, mutations that confer to the cell invasive capacity, the metastatic phenotype and resistance to therapy.

Finally, 'progression' refers to the terminal, presumably irreversible stage, in which the tumour demonstrates its metastatic properties.

These qualitatively different stages, from precancerous stages to stages that are increasingly more invasive and eventually metastatic, were introduced with the Somatic Mutation Theory and quickly became the standard way of partitioning the dynamic process, cancer.

2.3 The Clonal Genetic Model of Cancer

As we have seen, in the 1970s and 1980s scientific programs were dominated by the focus on ONGs. The main causes of cancer were thus to be sought in *activating* mutations, i.e., mutations that actively promoted cell proliferation.

Soon, however, Tumour Suppressor Genes (TSGs) were discovered too. Experiments showed a dominant behaviour of normal mouse cells over cancer cells when the two cell types were merged (Harris 1971), suggesting that the former had to carry genes that were opposed to tumourigenesis, thus showing a tumour-suppressor function. This and other data (Steel and Harris 1989; Harris et al. 1996) suggested that carcinogenesis does not require an *acquisition* of function (an assumption that was behind the definition of oncogenes), but rather the *loss* of it, perhaps through damage of some pattern of cellular differentiation (cf. Sect. 1.3 on this point as well). The ONGs are genes that mainly act in a dominant manner and whose overexpression causes an acquisition of function, leading to increased proliferative capacity and adaptation to the environment compared to other cells, including the progenitor cell.[6] In contrast, when the TSGs are mutated, such mutations

[6] Notice, however, that the relationship between ONG and their function is not 1:1: these genes usually integrate multiple pathways and are involved in various cellular functions at the same time, so that altering the gene's sequence does not necessarily alter functions (cf. Sonnenschein and Soto 1999 on this point). It is therefore most appropriate to say that ONGs are involved in crucial signal transduction cascades. For example, an external growth factor signal leads to cytoplasmic signalling to a receptor through the nuclear membrane transcription that activates a transcription factor. When a genetic change in the growth factor or its receptor occurs, it can easily lead to its constitutive activation, and this will alter the function and the expression of a number of other genes or of their protein products downstream. The neoplastic phenotype thus observed is the result of this

lead to a loss of function By logical necessity, their effect is recessive.[7] The acceptance of TSGs was not so smooth in the scientific community. Their study, which is conceptually very simple, technically called for a major research effort to demonstrate that the dominance of oncogenes was not the general rule. This recognition laid the foundations of subsequent theories of the cancer cell. As Weinberg (2014, p. 269) narrates, "a new skirmish broke out [about] which classes of mutant genes were really important for cell transformation: oncogenes or the tumor suppressor genes? A vocal advocate of tumor suppressor genes—indeed a founder of this field—dismissed the oncogene gold rush of the mid-1980s as an act of lunacy, a band-wagon effect". Beyond caricatures, the point of view of the reductionist is characterized by the hope "that a small number of molecular events might explain

---

process. A concrete example is c-Myc. The c-Myc ONG is implicated in the control of neoplastic proliferation and also in the control of cell differentiation, and can be activated through gene amplification, that is, small chromosome pieces containing many copies of the same gene. It can also be deregulated through translocations that involve the immunoglobin heavy chain gene (Silva et al 2005; Klein and Klein 1986). Mainly, there are different chromosomal rearrangements that can constitutively activate genes like c-Myc. In chronic myelogenous leukaemia, instead, the translocation (9; 22) leads to the formation of the Philadelphia chromosome. On chromosome 22, a hybrid gene bcr/abl is formed in which the abl gene, which contains an internal tyrosine kinase domain, is removed from its physiological control and is put under the same control of bcr. In this way, a constitutively active tyrosine kinase is produced. Another example is the ONG fins, which loses its ability to be inactivated by a single point mutation, with the result that the encoded receptor is constitutively active and is not responsive to negative feedback control.

[7]A typical example is the gene involved in retinoblastoma. Cavanei and colleagues located the gene for retinoblastoma (known as Rb) in a region of chromosome 13 (Cavenee et al. 1983). They suggested that both sporadic and hereditary type tumours were due to a second alteration that involves this gene, either through a new mutation on the second allele, or through Loss of Heterozygosis (LOH), i.e., the loss of the second normal gene through deletion or other major rearrangements on chromosome 13. LOH reduces Rb to a homozygous state so that the mutation on the first allele is finally responsible for retinoblastoma clinical manifestation. This characterization of Rb alterations to get the kind of cancer known as retinoblastoma confirmed the hypothesis dating back to Knudson (Knudson 1971). Studies of cDNA were important in this development. cDNA is a DNA molecule made as a copy of messenger RNA and therefore lacking the introns that are present in genomic DNA. cDNA clones represent DNA cloned from cDNA and a collection of such clones, usually representing the genes expressed in a particular cell type or tissue, is a cDNA library. In 1986, Friend and colleagues isolated the cDNA that mapped in the human Rb locus, and demonstrated that it is often deleted in a high percentage of tumours. Other groups, working with cDNA fragments from hybrid-transcripts of normal tissue compartments, discovered that this gene was abnormally expressed or deleted in all retinoblastomas. The experimental data confirmed the inactivation of Rb as a cause of this cancer as well (Lee et al. 1987; Friend et al. 1986; Fung et al. 1987; Huang et al. 1988). In familial forms of cancer, then, a mutant allele is transmitted either from the mother or father, while the second mutation affecting the Rb locus occurs in retinal tissue after birth, giving rise to retinoblastoma. Conversely, in the much rarer sporadic form, both mutations in the Rb locus are acquired by independent mutational events after birth. Therefore, while inheriting the abnormal allele determines a high incidence of tumour formation, in the case of sporadic tumours the TSGs have a low correlation with the onset of cancer, as demonstrated also in breast cancer linked to BRCA1. Even if new experimental evidence subsequently showed how even benign tumours could correlate with LOH (Harris 2005), this persisted as one of the diagnostic factors that are used especially for the identification of predisposition of hereditary tumours.

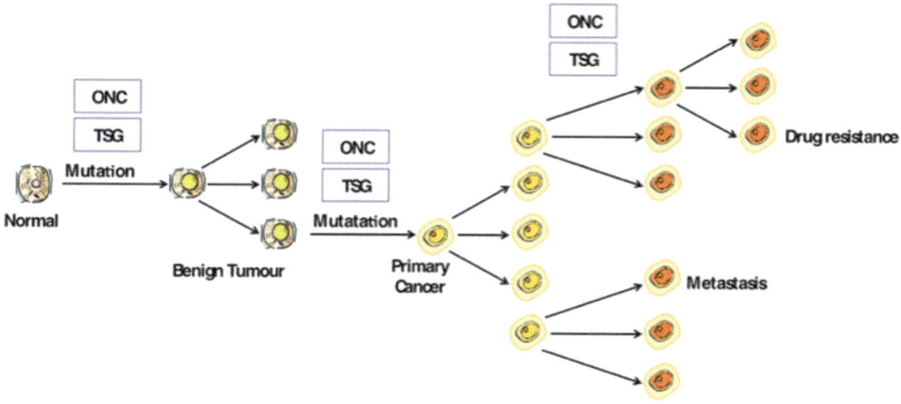

Fig. 2.1 Clonal genetic model (Modified from Feinberg et al. 2006)

cancer", so that the multiplication of causes is not welcome.[8] "As it turned out – Weinberg continues – the importance of both classes of genes soon became apparent. This notion acquired traction from the Vogelstein work of 1989 [...]. The greater the degree of progression a tumor exhibited, the larger were the number of somatic mutations affecting both oncogenes and tumor suppressor genes" (*ibidem*).

Eventually, knowledge of ONGs and TSGs was integrated within a unified picture: the Clonal Genetic Model of cancer. In this model, a normal cell undergoes a mutation in an ONG or in a TSG that usually requires another mutation to fully unfold its effects. TSG mutations are recessive mutations. ONG mutations, instead, can be either dominant or recessive, depending on the physiological function they compromise. At its simplest, the model predicts the functional impairment of these genes as the origin of a tumour cell's development, by clonal reproduction into the cells that constitute the tumour mass and ultimately trigger metastases. Feinberg et al. (2006) summarize this view of cancer in a critical review: cancer "arises through a series of mutations, including dominantly acting oncogenes (ONG) and recessively acting tumour-suppressor genes (TSG). Each mutation leads to the selective overgrowth of a monoclonal population of tumour cells, and each significant tumour property (invasiveness, metastasis and drug resistance) is accounted for by such a mutation" (Feinberg et al. 2006, p. 23) (Fig. 2.1).

In brief, that proto-oncogenes can be activated by retroviruses or somatic mutations was considered an explanation for the neoplastic phenotype. As reviewed also in Weinberg (2006), cancerous mutations could be of two types, identifiable as *gain-of-function* and *loss-of-function*:

[8]Weinberg even recalls emotionally that "For a brief moment in 1982, there was the illusion that cancer was as simple as it possibly could be—a normal cell differed from its neoplastic counterpart by one base out of three billion!" (Weinberg 2014, p. 269). He refers to a DNA sequencing study that revealed that the bladder carcinoma oncogene differed from its normal proto-oncogene counter-part by a single point mutation.

1. mutations that are related to a *gain-of-function* promote cell proliferation and generally correspond to the so-called ONGs that encode growth factors and their receptors, i.e., proteins involved in signal transduction pathways, mechanisms of the cell cycle and other similar processes;
2. mutations that are related to a *loss-of-function* inactivate inhibitory signals encoded by anti-oncogenes or TSGs.

2.4 The Stochastic Model of Cancer

In addition to the idea that cancer had a genetic basis, during the 1970s some experimental evidence emerged suggesting that cells underwent a progressive accumulation of mutations through different stages, in a stochastic manner. In 1976, Peter Nowell (Nowell 1976) incorporated in the existing models the idea that cancer was produced by multiple events, or "hits". For Nowell, tumour development and progression was mediated by the accumulation of, and selection for, genetic changes. The model was called the "Stochastic Genetic Model" (see Shackleton et al. 2009; Magee et al. 2012). Nowell argued that a genomically unstable, expanding population of cells could accumulate rare variants, leading to additional selective advantage of certain cells. Progressive cycles of mutation and selection would increase the number of clones that would dominate the cell population with an aggressive tumour phenotype. Support for this account came also from the observation that aneuploidy (i.e., abnormal number of chromosomes) is higher in solid tumours than in initial tumours, and that specific chromosomal alterations occur during the clinical progression of leukaemia. Further analysis of the mechanisms involved in the early stages of carcinogenesis revealed that defects in DNA repair or mitotic errors provoked by many cancer causing agents (e.g. ionizing radiation, viruses) could also induce potentially mutagenic changes that might increase subsequent tumour progression speed.

In the late 1970s, few mutations had actually been described in molecular terms. The well-known Philadelphia chromosome (Nowell and Hungerford 1960) was one of the very few specific large-scale translocations yet reported. Many DNA alterations are normally present that cannot be considered as mutations in functional terms, as they do not compromise the physiological activity of the affected gene. Such alterations are nevertheless stabilized over time, allowing for subsequent genetic drift, a phenomenon which is well known and widely studied in other fields of biological sciences. These elements left the door open to further research for the identification of the mechanisms involved in the initiation and progression of cancer. Many resources were invested in this enterprise. At the same time, it became clear that the epistemological issue of the functional definition of genes and their

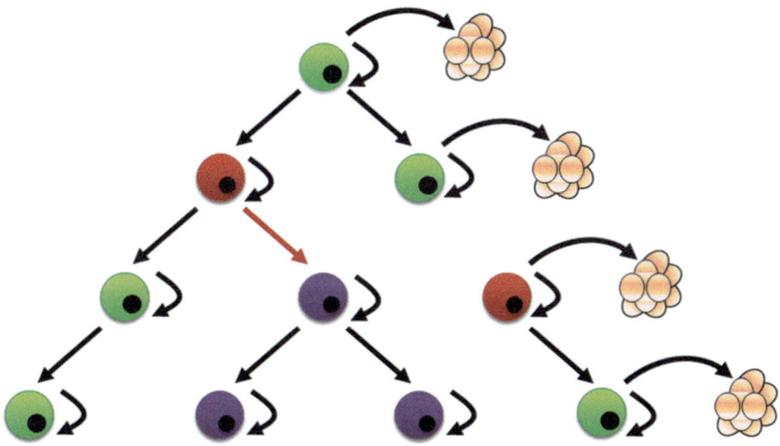

Fig. 2.2 The stochastic model (Modified from Vescovi et al. 2006)

relationship with the overall physiological or pathological behaviour of cells and tissues could not be overlooked (Fig. 2.2).

2.5 The Multigenic Multiphasic Model of Cancer

Empirical research persisted in the search for new genes or, alternatively, in the identification of interactions among known genes. Genes were still held to be responsible *in vitro* for neoplastic transformation, and *in vivo* for tumour onset. Then Bert Vogelstein illustrated how different kinds of alterations (in both ONGs and TSGs) were necessary, related and connected in the carcinogenesis of colorectal cancer. Fearon and Vogelstein presented their results in a landmark article in the history of cancer research. Their coherent and sequential tumourigenic process consisting of a series of stages of molecular events, the "Multigenic and Multiphasic Model of Cancer" (Fearon and Vogelstein 1990), reinforced the idea of the clonal evolution of tumour cells. This model was widely accepted by the scientific community and provided the scientific basis for initiation, promotion, transformation and progression, which until then were based on purely observational hypotheses. What was new in this model was the emphasis on the fact that, as these authors argued, the total accumulation of molecular changes was more important for tumour progression than their sequence or even identity, concluding that five or more genetic alterations were probably required for the development of carcinomas, while a lesser number of mutations would be required for benign tumourigenesis (Vogelstein and Kinzler 2004).

Vogelstein also categorized tumour genes in a new way in a paper published in 1997 (Kinzler and Vogelstein 1997). ONGs and TSGs were classified in:

- *Caretaker genes* that participate in the maintenance of the integrity of the genome and enable the cell to transmit an identical genome during successive cell divisions; they act as the guardians and curators of the genome; they encode products that stabilize the genome. Mutations in these genes lead to genomic instability. Two distinct classes of genomic instability give rise to tumour cells: mutational instability, leading to changes in the nucleotide sequence of DNA, and chromosomal instability, provoking improper segregation of chromosomes (see also van Gent et al. 2001).
- *Gatekeeper genes* regulate cellular homeostasis and the cell cycle by controlling the entry of the cell into the various phases of the cell cycle. In contrast to caretaker genes, gatekeepers encode gene products that prevent the growth of potential cancer cells and block the accumulation of mutations that would lead to increased cellular proliferation (see also Frank 2003; Campisi 2005).

This new classification of genes on the basis of functional categories, outlined in the Multigenic and Multiphasic Model, would later on be conserved in the epigenetic model of cancer.

2.6 The Epigenetic Progenitor Model of Cancer

While mutations and genes were progressively associated with disease progression stages, little by little a multitude of changes that were not strictly genetic, but rather epigenetic alterations of genes' expression, were documented. Epigenetic factors regulate gene expression, while not being related to specific changes in the DNA sequence. One of the first associations that was noticed between neoplastic phenotype aberrations and epigenetic mechanisms was the correlation between abnormal DNA methylation and increased chromosome instability (Feinberg and Vogelstein 1983).[9] Both the methylation of DNA and histone acetylation are involved in the normal processes of development as well as in chromatin organization and remodelling of nucleosomes, as in some forms of loss of imprinting,[10] originally reported in

[9] In the wake of the Multistep Model, in recent years, a number of genes have been identified showing hypo-methylated DNA, usually in the promoter in pre-invasive stages of colon cancer and other cancers, but which are rarely mutated. These genes have been named "Epigenetic Gatekeepers", assuming that their normal operation was to prevent a cell from acquiring an immortal phenotype or the ability of self-renewal, typical of malignant phenotypes, through epigenetic regulation (Jones and Baylin 2007). This possibility would be consistent with the presence of dysplastic areas that appear in the gut epithelium before a benign tumour is clinically detectable, that are not attributable to alterations in the genome sequence but rather to those in the epigenetic program regulating differentiation of the stem cell compartment of that tissue.

[10] Genomic Imprinting is the situation where a gene is either expressed or not expressed in the embryo depending on which parent it is inherited from.

relation to hereditary cancers like Wilms' tumour where the expression of a normally silent allele was activated (Scrable et al. 1989). Speculation began that these mechanisms could play an important role in the beginning of the neoplastic process.[11]

Some epigenetic alterations were typically seen in cells at early stages of transformation. This observation provided good evidence for a causal-mechanistic scheme to justify the causal relevance of epigenetic factors and to integrate them into the existing genetic models (see Feinberg et al. 2006). The so-obtained "Epigenetic Progenitor Model of Cancer" proposed that a key step in tumour formation – before the process is mediated by the accumulation of genetic mutations – consists in the epigenetic alteration of pre-cancerous cells. Cancer would thus arise in three stages (Feinberg et al. 2006):

1. First, epigenetic alteration affects a progenitor (or stem) cell in a certain tissue, mediated by aberrant regulation of TSGs. This alteration may be due to events in the cell itself, to the influence of the stromal compartment, to environmental damage or to other types of insult.
2. The second step involves mutation: a mutation in a gatekeeper gene (GKM) or in a TSG in solid tumours, or rearrangements of ONGs in leukaemia and lymphomas. Although these GKMs are themselves monoclonal, the progenitor cell expansion increases both the risk of epigenetic cancer and the frequency with which primary tumours can occur separately. Epigenetic alterations can also enact an ONG-induced mutation or the silencing of a TSG.
3. The third step is genetic and epigenetic instability leading to tumour evolution. Many of the properties that favour the tumour (invasion, metastasis, drug resistance) are typical properties of the progenitor cell that gives rise to the tumour and do not require other mutations, further underscoring the importance of epigenetic factors in tumour progression (Fig. 2.3).

Experimental data has repeatedly highlighted that the transformation of a normal cell into a *precancerous state*, and the reversion from a malignant to a normal phenotype, occur with a frequency that is hardly compatible with mechanisms that involve only DNA damage or its repair. The same can be said of the amount of time, usually months or years, between the formation of a lesion and the occurrence of

---

[11] Indeed, epigenetic alterations in gene expression that persist after exposure to carcinogenic chemicals have been increasingly identified as important factors in the initiation and progression of cancer (Feinberg and Tycko 2004; Fukushima et al. 2005). As recent studies have revealed, the silencing action of many TSGs was indeed physiologically mediated by mechanisms such as hypermethylation of CpG islands in the genes' region that regulate its expression, the 'promoter', or as the recruitment of transcription factors or inhibitors of enzymes such as histone deacetylases, which produce functional modifications on histones, and the Methyl-Binding Protein (Jones and Baylin 2007). Although some chemicals that cause cancer in rodents are not themselves genotoxic carcinogens in humans, both genotoxic and non-genotoxic compounds were described as being able to alter, at some stage of the neoplastic process, gene expression, via the induction of DNA transcription through methylation of histones, or other nuclear mechanisms, that influence the activity of the transcriptome, without the occurrence of genetic mutation (Jones and Baylin 2007).

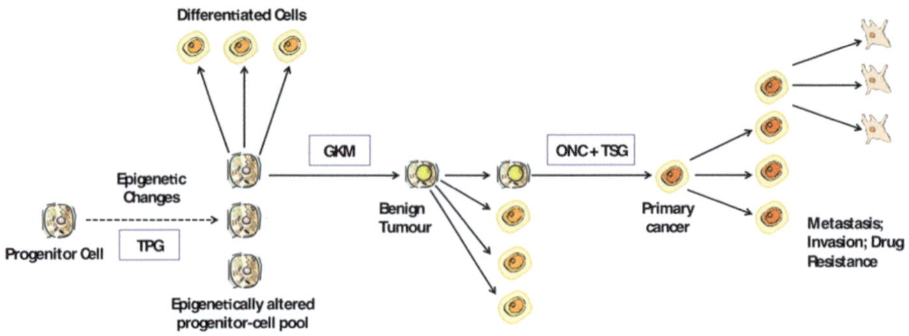

Fig. 2.3 Representation of the epigenetic progenitor model (Modified from Feinberg et al. 2006)

oncogenic mutations detectable by PCR (i.e., tumour latency, Sect. 1.3.1). On the other hand, the expression of proto-oncogenes detected within minutes or hours after a critical event for the normal genome is easily compatible with changes in gene expression mediated by epigenetic events, which would then be more easily correlated with the first steps of the neoplastic process (Jaffe 2005). Other empirical observations which highlight a possible causative role of epigenetic events on the onset of cancer were the following (Feinberg 2007): (1) epigenetic changes precede the onset of cancer and confer a higher risk for developing neoplasms, (2) hypomethylation of DNA also confers an elevated risk, (3) the increase in genomic instability that seems crucial in the early stages of tumour progression occurs also in response to treatment with carcinogens. In recent years, several lines of experimental evidence have supported the validity of the Epigenetic Progenitor Model (Feinberg et al. 2006). To start with:

- Many cases of the reversibility of the tumour phenotype described in literature have been demonstrated for leukaemia as well as for solid tumours;
- Nuclei of mouse melanoma cells can differentiate in inbred normal mice, suggesting that tumour cells can be reprogrammed in view of normal development, that they are controlled epigenetically, although not completely, as these clones show a higher incidence of melanoma compared to normal cells; mutations seem to act as stable genetic predisposing factors;
- Serial transplantation of cancer highlights how the daughter cells maintain expression, at different levels, of molecular markers of the original primary tumour. These data and empirical potentially related empirical evidence were indicative not only of the Epigenetic Progenitor Model, but also of the fact that cells possessing some stem cell properties, like self-renewal and differentiation are present in tumour tissues.

The Epigenetic Progenitor Model also accounts for some aspects of tumour heterogeneity (1.4.2, 5.2) and for the dependence of cancer onset on environmental risk factors better than any purely genetic model. Epigenetic factors can thus be the first step of neoplastic progression, destroying the normal program of a progenitor or stem cell from which the tumour will arise, for example by stimulating cell proliferation outside the normal microenvironment (Feinberg et al. 2006).

Interestingly, inherent to this model is the concept of the *epigenetic field for cancerization*. Originally introduced for the oral cavity, to identify mucous membranes that are in some way predisposed to the onset of cancer (Slaughter et al. 1953), this concept describes the clear relationship that exists between the levels of methylation – for example in the gastric mucus, although there is still no histological evidence of malignant changes – and the risk of developing cancers. The epigenetic field for cancerization is thus able to induce genomic instability, hypo-methylation and tumour progression, as well as aberrant transcription of several genes (Eden et al. 2003). But subsequently similar areas were described in other organs such as the stomach, upper digestive tract in smokers, the oesophagus in smokers and alcoholics and the bladder (Ushijima 2007).

In the next chapters we will see how important the notion of "field" is from an epistemological point of view. With the notion of epigenetic field, epigenetic alteration and modulation determine functionally relevant modifications that are triggered by contextual factors. The microhabitat, surrounding the progenitor cells, contains informational factors to which epigenetic features of the cells are linked. In turn, epigenetic features regulate genes' expression over time with no change in the nucleotide sequence.[12] A level different from the genetic one is thus explanatorily relevant with respect to cancer origin. Such level is characterized by its regulatory role in the processes of cell differentiation.

2.7 The Hierarchical Model of Cancer

Not all cells of a primary tumour are able to proliferate and form a colony *in vitro* (Hamburger and Salmon 1977) and many tumoural forms (like, for example, teratocarcinomas) have a cytoarchitecture reminiscent of embryonic morphologies (Clarke et al. 2006). Tumour cells are functionally heterogeneous: only a small fraction of cells within each tumour carries a tumourigenic potential *in vivo*, when transplanted into immunodeficient mice (Bruce and van der Gaag 1963). Cells within a cancer apparently correspond to different stages of development. This was assessed, for example, in experiments on epithelial tumours: the majority of tumours

[12] Epigenetics seems to be at the heart of all developmental processes of cellular differentiation and proliferation. It thus provides crucial data in order to understand what mechanisms, for example, make stem cell maintenance and differentiation, or ageing and cancer processes, possible. Dealing with *processes* and *signals*, epigenetic control unravels a regulatory program, which has been questioning our understanding of systemic control (see Chaps. 3, 4 and 5). Interestingly, it is difficult to define epigenetics through a positive statement (i.e. saying what it is). We usually describe epigenetic mechanisms saying that they *are not* related with alterations on the genetic sequence. This gives way to the reflection on the epistemological and ontological status of regulatory processes in biological systems and on the explanatory relevance of the context in biological explanations (cf. Bertolaso 2013b, Chapter 5).

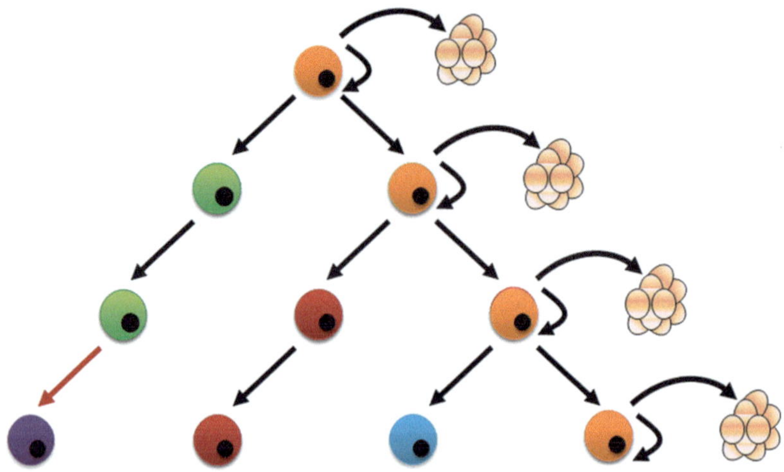

Fig. 2.4 Representation of the hierarchical model (Modified from Vescovi et al. 2006)

seemed to contain a heterogeneous population of fully or partially differentiated cells, which mirrored that of normal organs (Pierce et al. 1977). This kind of empirical evidence laid the foundation for the theory of the Cancer Stem Cells (CSCs) and for the "Hierarchical Model of Cancer" that considers the CSCs as promoting the neoplastic growth of a tumour. In this currently highly considered model, cancer does not arise, as maintained by the SMT, by a mutation in a somatic cell which results in increased proliferation, but rather by modifications in a few cells that persist for a long time in the tissues. Cancer might be considered, in the first place, as an adjustment of some biological characteristics of aberrant stem cells and their hierarchical organization. The properties of CSCs can be summarized in their capacity for self-renewal and differentiation, in their proliferative capacity throughout the life of the host, in their scarce number in normal and tumour tissues and in their quiescent or low proliferative rate. Their primary role in tissue renewal seems to match well with the analogies mentioned in Sect. 1.2.1 with other phenomena, such as embryonic development and regeneration after injury or lesions of various types (Boman and Wicha 2008) (Fig. 2.4).

For some authors, the Hierarchical Model also explains the failure of some therapies (such as chemotherapy and radiation therapy) that act on actively proliferating and differentiating cells. While the proliferating offspring of CSC, in fact, are destroyed by some therapies, their quiescent counterpart is not, ready to start again with a new cycle of clonal expansion later on (Boman and Wicha 2008). CSCs, by virtue of their characteristics, would thus be responsible for the drug resistance frequently observed in tumours. The rate of mutations and tumour cell differentiation is also off the control of drug therapies aimed at the elimination of actively proliferating cells.

The existence of CSCs is a subject of debate within the medical research community, because many studies have not been successful in discovering similarities and differences between normal tissue stem cells and cancer (stem) cells, although conclusive evidence for CSCs was published in 1997 Bonnet and Dick. In any event, the use of the concept of CSC is justified by the analog clonogenic capacity of tumour cells and of their normal stem cells counterpart. *Self-renewal* and *differentiation* into various cell lines activates the expression of telomerase, apoptotic pathways, membrane transport activity and higher ability to migrate and metastasize (Clarke et al. 2006).

The Hierarchical Model views cancer not as a simple clonal expansion of transformed cells, but as a complex three-dimensional tissue in which cells become functionally heterogeneous as a result of organogenic differentiation (Dalerba et al. 2007). The structural and functional heterogeneity of tumours no longer needs to be explained by postulating the coexistence of genetic sub clones, created by the gradual accumulation of independent somatic mutations, and under changes in the tumour microenvironment. Tumours appear to function as complex organs that have undergone an aberrant development and act by caricaturising tissues' normal growth promoted by a pathological counterpart of normal adult stem cells, the CSCs. CSCs, like normal stem cells, give rise to a hierarchical organization of cell populations that underlie organogenesis (Reya et al. 2001; Pardal et al. 2003).

More specific assumptions and predictions of the Hierarchical Model are:

(a) Stem cells that survive a long time in tissues are more exposed to accumulate mutations that lead to cancer. This is plausible because the default state of this cell type should be quiescence or slow prolificacy, with a cell cycle that is in average longer than that of other cells;
(b) The daughter cells inherit mutations from these stem cells, providing an area more prone to the transformation events of the typical malignancy;
(c) Tumours have a hierarchy of cells that is a caricature of normal ontogeny because they reflect, in some way, the normal pluripotency of the original cell;
(d) The deregulated pathways in tumours are those that are involved in the development of various organs during embryogenesis (sonic hedgehog, Notch, PTEN, BMI-1, WNT, and p53; see Boman and Wicha 2008; Lobo et al. 2007).

Epigenetics also came to be important in the Hierarchical Model. Evidence of ubiquitous epigenetic changes in cancer (hypo-methylation, LOH,[13] etc.) as well as the presence of progenitor cells in normal tissues of patients with tumours (Feinberg 2007), led to the establishment of a relationship between the Epigenetic Progenitor Model (see Sect. 2.6) and the CSC hypothesis. Incidentally, the Hierarchical model, like the Epigenetic one, supports the hypothesis that cancer has a polyclonal (not monoclonal, as assumed by the SMT) origin. Unlike the Clonal Model, the Hierarchical Model of cancer "implies that only a small subpopulation of tumour stem cells can proliferate extensively and sustain the growth and progression of a neoplastic clone" (Vescovi et al. 2006, p. 427).

The Hierarchical Model of cancer, integrating the epigenetic one, established a new coupling between the timing sequence of neoplastic progression and the concept of cell differentiation and tumour heterogeneity: a (dis)organization of differentiated units accounts for cancer heterogeneity and its temporal dynamics, i.e.

[13] See footnote 7 for an explanation of LOH.

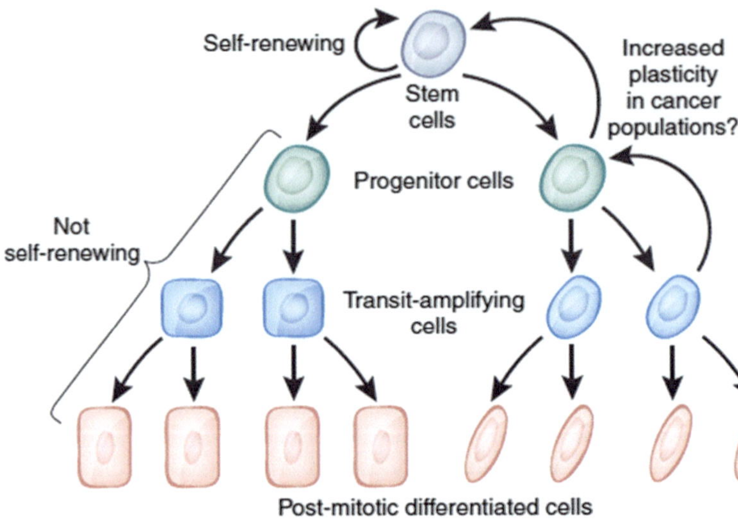

Fig. 2.5 Stem-differentiation hierarchy. The different dimensions of tumorigenic proliferation of CSC are here represented: self-renewing, (re)production of tumor progenitor cells, and production of (aberrant) differentiated cells. The problem is shown that increased plasticity in cancer populations could enable bidirectional interconvertibility between CSCs and non-CSCs. Reproduced from Gupta et al. (2009)

different stages of differentiation, some of which retain the tumourigenic properties; a small subpopulation of tumour cells proliferate extensively and sustain the growth and progression of a neoplastic clone. Notice that in this way CSCs became explanatory of tumour initiation *and* tumour growth. But the production of differentiated non-tumourigenic offspring by those cells (Vermeulen et al. 2008) may seem contradictory. Indeed, logically speaking, it is. Given the properties of a stem cell and tumorigenity, the proliferation of CSCs could be supposed to proceed in either direction (CSC<-->TC) at the same rate, making any link between CSC and oriented division and functional behavior (Fig. 2.5) logically untenable.

More will be said in Chap. 7 about the Hierarchical Model. We will see how a theory of biological explanation (outlined in Chap. 6) can describe the passage from clonal models to the CSC hypothesis, and make epistemological sense of the problems encountered.

2.8 The Evolutionary Argument

The Editorial of the March 2009 issue of *Nature Collection on Cancer* begins: "*The development of cancer is an evolutionary process* that is driven by multiple genetic and epigenetic changes" (my emphasis). Indeed, most of the theories and cellular models of cancer have been expanding at some point their explanatory accounts into an Evolutionary Somatic Model of cancer, which implies the development of the disease by progressive (natural) selection of the most malignant cells along with the progressive accumulation of mutations in TSGs and ONGs. This evolutionary hypothesis, in its simplest formulation, states that mutations that result in overgrowth of a monoclonal population of tumour cells undergo positive selection. A mutation of an ONG or TSG is followed by the expansion of a benign tumour; additional mutations lead to the primary tumour, then to its expansion, through the loss of the genomic integrity of the cells, and ultimately to tumour transformation, from benign to malignant.

As we have seen in Sect. 2.4, the Stochastic Model of cancer is conducive to this perspective: the more aggressive neoplastic phenotype is represented in various sub-clones that are formed by the first selective process and have peculiar characteristics as, for example, drug resistance and the ability of giving rise to metastases. In the Hierarchical Model (Sect. 2.7), comprehensive of epigenetics,[14] Cancer Stem Cells (CSCs) are identified, for their ability to undergo clonal proliferation, as the privileged target of the selective pressure that leads to the development of cancer.

Somatic evolution exploits the accumulation of mutations in the cells of the body (soma) during its life cycle, and the effects of these mutations on survival and reproduction (fitness) in the cells. In fact, the mutations that accumulate stochastically within the expanding populations of clones occur not only at the level of ONGs and TSGs, but may also involve other genes. The main idea is that the cells in a pre-malignant state become malignant by evolving through natural selection (Nowell 1976; Merlo et al. 2006) and that this phenomenon is shared by other physiological processes such as ageing.[15]

Lewontin (1970) formalized three necessary and sufficient conditions for natural selection that became classic: (a) variation, (b) inheritance, and (c) fitness. According to the evolutionary argument, somatic cells can undergo natural selection because (a) they are arranged in local populations that exhibit variation, (b) they transmit their genetic and epigenetic features to daughter cells, and (c) their variations may affect their persistence and proliferation, conferring relative selective advantages. Hanahan and Weinberg, moving along these same premises, suggested in their article, "The Hallmarks of Cancer" (Hanahan and Weinberg 2000), that cancer can be described by a small number of functional ingredients, despite the complexity of the pathology. Tumour progression would proceed according to a process that is similar

[14] Epigenetic changes are a substitute for these mutations or genetic alterations, in the sense that they affect genes' effects, as a mutation would do.

[15] As Silvia Caianiello pointed out (personal communication), this position would have the 'metaphysical consequence' to give ontological consistency to the 'bad', creating a new caricature, a parody of natural selection. I think that this is not necessarily the case. It would be should we consider the Evolutionary Argument as a model, and, furthermore, as a satisfactory explanation in its own right. As will be clearer in the next chapters, however, this is not the case in some research programmes. The point is that scientists are often not able to make the conditions of validity of their own models explicit. As I will argue in Chaps. 5 and 6, peculiarities of biological behaviours do challenge epistemological assumptions as well as the domain of validity of models.

to Darwinian evolution, where each genetic change confers a selective advantage for cell growth and where genetic instability, a common feature in many cancers, would constitute an "enabling characteristic" that facilitates the acquisition of additional mutations due to the damage that a cell could have previously undergone to its DNA repair system. The six functional characteristics attributed to genetic alterations spelled out by Hanahan and Weinberg are:

1. "Self-sufficiency in growth signals": this refers to the observation that tumour cells produce their own growth factors, not depending on external growth signals.
2. Insensitivity to antigrowth signals: normal cells are maintained in a state of quiescence by growth inhibitory signals; genetic changes confer tumour cells the ability to ignore these signals.
3. Evasion of apoptosis: normal cells will activate the self-destruction or apoptosis program in response to irreversible DNA damage, insufficient growth signals or ONG overexpression, while tumours acquire the means to "evade apoptosis", resulting in an accumulation of altered cells.
4. "Limitless replicative potential": the majority of mammalian cells generally proliferate for a limited number of times due to progressive shortening of the chromosome ends, or telomeres; virtually all malignant cells have acquired the ability to maintain their telomeres.
5. "Sustained angiogenesis": cancer cells promote the formation of blood vessels; this is essential for the tumour, because cells cannot survive at a distance of more than about 100 uM from blood vessels.
6. Ability to branch out through invasion and metastasis: during the development of most tumours, primary cancer cells acquire the ability to spread from inside the surrounding tissues to distant sites, giving rise to secondary tumours in remote organs.

If these genetically determined functional characteristics describe the malignant phenotype, the pathways that cells undertake to arrive at a malignant phenotype, however, are variable, and the order in which the "hallmarks of cancer" manifest themselves can change from tumour to tumour. The early events of tumourigenesis are difficult to measure and identify clinically but, assuming a somatic evolution of cancer, they can be simulated in accordance with other principles such as those of biological evolution (Spencer et al. 2006). Therefore, tumour cells compete for resources, like oxygen and glucose, and for space. A cell that acquires a mutation that increases its fitness will generate more daughter cells than its non-mutant competitors. In this way, a mutated cell will form a clonal, expanding population, thus manifesting the characteristic signature of the natural selection of cancer.

Metastasis would thus be the endpoint of a long process of selection – the change over time in a population of cells in a tumour due to heritable differences that make a difference to relative survivorship and reproductive success. But why and how exactly does metastasis evolve? One explanation asserts that metastasis does not require new mutations, but is linked to the fact that cancer cells take control of complex biological programs, normally involved in the maintenance of cellular and organ-related physiological processes (Weinberg 2008). This is the case of the mechanisms of the Epithelial-Mesenchymal Transition (EMT) playing important roles in normal morphogenesis (Thiery 2002). An EMT is a process that allows a polarized epithelial cell, which normally interacts with basement membrane via its basal surface, to undergo multiple biochemical changes that enable it to acquire a mesenchymal cell phenotype, which includes enhanced migratory capacity, invasiveness, elevated resistance to apoptosis, and greatly increased production of ECM components. In tumour cells, these processes would be used in an aberrant way, allowing the cells to assume an invasive phenotype. The dissemination of tumour cells in the organism, therefore, takes place through a cascade of subsequent events (Fidler 2003) in a process called "colonization", that ends up with the metastatic invasion of new organs and tissues.

There are also data indicating that a metastatic phenotype, more than being a property acquired during the process, would already in some way be present in the cells of the primary tumour. Analyses of the expression of a large number of well-characterized genes in tumours evidenced that some metastatic tumour cells express the genes in a manner much more similar to the primary tumours from which they originate than to the other tumours that were induced simultaneously. Tumour cells also metastasize to specific organs, suggesting that the differentiation program of the progenitor cell, which is context dependent, also constitutes a strong determinant of metastatic dissemination. Finally, the sets of genetic lesions present in metastatic cells resemble those present in the cells of the primary tumour (Weinberg 1988; 2008).[16]

In view of the somatic evolution hypothesis of cancer presented above, at the cellular level there is selection for those cells that manifest greater survival and proliferative capacities, for which a mutated cell that acquires all six "hallmarks of cancer" will be more competitive than the others that have not (yet) acquired them all; at the level of the organism, cancer is instead usually lethal, so that it is plausible that a number of genes and tissue organization mechanisms that suppress cancer might have been developed (Pepper et al. 2007; Cairns 1975). Therefore, it has been said that the neoplastic phenomenon is an example of what evolutionary biologists call "multilevel selection". To what extent this theory is tenable will be further discussed in Chap. 7 from the point of view of functional attributions to molecular parts.

---

[16] Such evidence can be either interpreted as a proof of the irreversibility of some steps of cell differentiation or even as evidence in favour of the clonal origin of cancers that was assumed all through the present chapter. The next chapters will elaborate a more articulated view of cancer and of the role of clones and genetic mutations.

As Nowak (Nowak 2006 cited in Vineis et al. 2010) suggested, the Darwinian paradigm may become a unifying theory for several biologic phenomena.[17] In fact, a Darwinian account leaves a way open to reconcile different cellular models of cancer. A model based on the concept of 'Darwinian' cell selection encompasses three models of cancer: model 1 is mainly centred around mutations, model 2 has to do with genome instability and deals with familiarity, while the main features of model 3 are instead clonal expansion and epigenetics, mirroring the classification I have been adopting up to now (Clonal, Stochastic and Epigenetic models of cancer).

Later in the book, we will see how systemic approaches deal with Darwinian selection through the notions of landscape and attractors (Sect. 3.3.2), and we will also critically reflect on the use of the evolutionary argument to justify the attribution of "functions" to cells (Sect. 7.2). I will use epistemological arguments to demonstrate that the evolutionary argument is, at present, often employed as an *ad-hoc* justification of aprioristic function attribution. A thorough consideration of the dynamics of living systems and of the epistemology of biology will show that the crucial issue for any evolutionary explanation resides in the proper consideration of its conditions of validity.

2.9 The Endurance of a Cell-Centred Perspective

We have seen how the fundamental assumption of the Somatic Mutation Theory – i.e., that cancer is the clonal expansion of cancerous cells due to genetic mutations – was progressively relaxed throughout decades of cancer research. Overcoming a mere hunt for cancer mutations, explanations of cancer modeled genetic mutations in a more and more complex way and combined them with epigenetics. Rigid monoclonality led the way to polyclonality, while the explanatory focus on the cell was contextualized by the Stochastic Model (Sect. 2.4) and the evolutionary argument (Sect. 2.3) and by the Hierarchical Model (Sect. 2.6). Throughout all these models, the importance of the cell and of its internal mechanisms remained firm.[18] In other words, cancer research remained largely within a perspective that I would like to call Cell-Centred Perspective. The evolution of Cell-Centred research programmes represent "the same attempt to find *common causative molecular patterns*

[17] Vineis et al. (2010) suggest that the term 'Darwinian' needs to be used cautiously, "being a short cut for 'somatic cellular selection'" (*ibidem*, 1703): it has entered into use in cancer literature, but "it should not be used to imply that Darwinian selection at the population (rather than cellular) level is involved in carcinogenesis" (*ibidem*, 1704).

[18] The emphasis on genetic mutation also remained very stable, albeit it was complexified, for example with the functional diversification of kinds of mutations that get combined in the Clonal Genetic Model (Sect. 2.3) and then complemented by epigenetics. In the Stochastic Model, cancer develops as a heterogeneous population of cells, not as the clonal expansion of one cell. Still, the events that differentiate the cells within a population of tumour cells are genetic in nature.

but (…) focused on different levels of genetic/epigenetic or cellular organization and their response under all kinds of environmental stress" (Heng et al. 2009, p. 538; my emphasis).

Even those models that pointed outside the cell were actually expressed in Cell-Centred terms or translated into Cell-Centred empirical research. Take the Stochastic Model (Sect. 2.4) as an example. The model could be seen as shifting the attention from the cell to a heterogeneous population of cells: cancer depends more on the dynamics at the level of cell population than on single genetic mutations. Yet, the tumourigenic potential still resides in cells. In fact, on the one hand, the stochastic model "proposes that tumour cells are heterogeneous"; on the other hand, it allows "that virtually all of them can function as a tumour-founding cell, although this might happen only rarely" (Vescovi et al. 2006, p. 427). In other words, if a meaningful explanation requires extending the focus from the cellular characters to the features of a population of cells, still, the population remains almost nothing but an aggregation of individual cells, all of them potentially tumourigenic by virtue of – once again – the mutations that occur in the context of stochastic dynamics. The Hierarchical Model, with its focus on CSCs, is equally Cell-Centred. In a Cell-Centred Perspective, carcinogenesis is conceived as a cellular-genetic phenomenon, a cell-autonomous process, intrinsically dependent upon the inherent properties of the cancer cell (Hanahan and Weinberg 2000). Carcinogenesis is defined in terms of errors in cellular proliferation, differentiation or (defect of) programmed cell death. In a Cell-Centred Perspective, the tumour cells are to be seen as normal cells 'gone mad' (Weinberg 2006): a tumour is formed when a cell of a metazoan undergoes aberrant changes in programs that normally regulate growth. This "one renegade cell" forms the tumour through its unregulated proliferative activity in spite of the organism's multiple levels of surveillance mechanisms to avoid this type of cell development (Weinberg 2006). The renegade cell resets its objectives endorsing only one: to create copies of itself.

The molecular circuitry that programs the transformation of the malignant cells became – and remained – the main objective of the research (Hanahan and Weinberg 2000). The desired outcome of this research programmes is described by the so-called "emerging integrated circuit" of the cell published in 2000 by Weinberg: an actual circuit of molecules that should provide a full explanation of the pathogenesis of cancer (Fig. 2.6).

According to Weinberg himself, this research programme is adequately described in terms of *reductionism*: "Reductionism has been the driving obsession of many biologists over the last decades. This obsession has carried us far through many of biology's thorniest problems, including the search for the origins of cancer" (Weinberg 1988, p. 1963). Progress in genetic discoveries regarding bacteria and yeast also contributed to the postulate that the cell genotype determines all aspects of the phenotype. As Weinberg emphasizes, when working with complex systems, the way to get rigorous conclusions is to reduce these systems to simpler subsystems and to study their components separately. The conclusions drawn, though

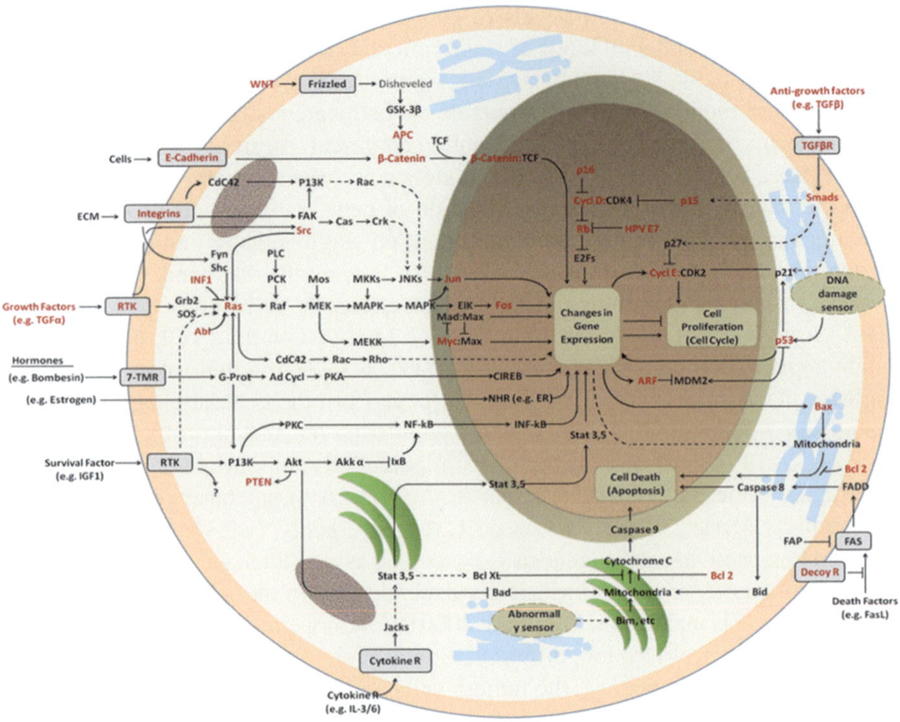

Fig. 2.6 The integrated cellular circuit (Modified from Hanahan and Weinberg 2000)

regarding and circumscribing only small portions of the much more complex system, are solid and permanent, and will not change, although they may be replaced by new perspectives in the work of future generations of researchers (Weinberg 2006). In cancer research, as we have seen, this has resulted basically in the characterization of genetic and epigenetic events involved in tumour progression.

Molecular biologists and biochemists were inclined to embrace such reductionist methodology and for many of them this meant taking a reductionist epistemological framework as the setting for research.[19] The 'reductionist pact' dominates with its fundamental and most popular explanatory assumptions: (1) that sooner or later all the attributes of cancer can be understood in terms of genes belonging to the tumour cells, (2) that all the functional properties of a tumour can be traced down directly to the behaviour of individual tumour cells within the tumour mass (Weinberg 2006). Although the second assumption got repeatedly questioned by clinical and molecular evidence regarding tumour heterogeneity, within a strong reductionist paradigm all steps of neoplastic progression could be traced back to a gene, or a small number of genes, i.e., "the root causes of cancer" (Weinberg 2006, vii) or other phenomena that can simulate the effects, such as epigenetic mechanisms.

[19] For more background on these terms, please see the Appendix and Bertolaso (2013a, b, c).

Reductionism does not truly coincide with embracing genetic determinism nor claiming that molecules are all that matters. Rather, any reductionism proceeds by adopting a mechanistic causal perspective through which the systems are decomposed (see Sects. 5.2.2 and 5.4, and the Appendix, for a discussion of mereology, i.e., the parts-whole organization and decomposition). According to Hanahan and Weinberg: "[C]ancer biology and treatment […] will become a science with a conceptual structure and logical coherence that rivals that of chemistry or physics" (Hanahan and Weinberg 2000, p. 67–68) and the construction of an integrated cell circuit, using the information that we accumulate over the coming decades, will provide the tools for this: "With holistic clarity of mechanism, cancer prognosis and treatment will become a rational science" (*ivi*, p. 67). The hope is that empirical research could transform itself more fully into a logic-based science to derive even complex pathologies from a few common (mechanistic) principles (*ibidem*). Virtually all mammalian cells possess the molecular machinery that carries out the regulation of proliferation, differentiation and cell death dictated, however, by specific genetic programs responsible for the features of the cells and ultimately the whole organism. The fundamental problem is then to identify the *mechanisms* that allow the genetic structure of a cell, and eventually of an organism, to determine phenotype and functions (Weinberg 2006).

On one hand, the kind of explanation characterizing the Cell-Centred Perspective can thus be labeled *mechanistic-reductionist*: it is based on the belief that any collective and macroscopic behaviour can be reduced, through appropriate strategies, to microscopic information about the constituents of the system; and it is focused on molecular mechanisms that, once put back together, will form the integrated regulatory circuit of the cell whose alteration is (the cause of) cancer. On the other hand, emphasis on the autonomy of tumour cells reveals a tendency to embrace *biological atomism*, i.e., "(t)he doctrine which postulates a basic indivisible unit of life and seeks to explain the physiological and morphological constitution of all living beings operation in terms of these fundamental units" (Nicholson 2010, p. 203).

2.10 Reductionism Against the Wall of Complication

The mechanistic-reductionist aspirations and atomistic inclination of the Cell-Centred Perspective (see above, Sect. 2.9) ran into complications after complications: first was the multiplicity of genes and networks involved in cancer, then the different mechanisms compromised in the progression of cancer, the importance of intracellular communication, and the role of stromal cells in the onset and maintenance of the neoplastic phenotype. From the Cell-Centred point of view, struggling to get a simpler and more comprehensive insight, cancer manifested itself as *complicated*. At the same time, the Cell-Centred Perspective integrates the discovered contextual factors into the explanatory models through an unchanging epistemological strategy: an additional element of an integrated circuitry growing huge. The tension is nicely summarized in this quote: "Cancers exist in an extraordinary variety of taxonomically, quasi-classes, genera, species, characterised by divergent cells of origin and mutational spectra. Each cancer is unique" (Weinberg 2006, p. 307). Nevertheless "[t]he number of mutations found in any cancer can vary from a handful (10–20) to (the more usual) hundreds or thousands" (*ibidem*) so that no necessary and sufficient conditions can be found in mechanistic terms. Every cancer is a disease in itself, to the point that molecular heterogeneity seems to be the distinctive characteristic of cancer. Moreover, the explanatory relevance of cellular components change over time depending on the different stages of carcinogenesis. Regression of the neoplastic phenotype and metastasis remains unsatisfactorily explained within the Cell-Centred Perspective.

Perhaps, at least in the mechanistic-reductionist explanatory strategy, the answer is that there are no causes at all. More plainly, the neoplastic disease is far from being an endless monologue of cancer cells that speak to themselves, although it is not clear what this actually implies from an explanatory point of view.

If genes cannot be considered the most fundamental explanatory elements of a complex process like carcinogenesis, where are the principles that, once compromised, are responsible for carcinogenesis? Why tumor cells do not always retain their neoplastic capability? Where are the dynamics to be described in order to explain why we do or do not develop cancer? What are the implications for our understanding of biological explanations and functional accounts in bio-medical sciences?

Weinberg (2014) declares that molecular biologists had entered cancer research in the 1970s hoping that they "would ride in—knights on white horses—and save the day" (p. 267). From time to time, key findings gave new hope that a genetic and Cell-Centred key to cancer would be found. Now that such hope has been frustrated, the problem for Weinberg is that research and technology have multiplied information and created a "data deluge". "We don't know how to integrate datasets – Weinberg writes – and we don't know how to deal with this complexity". While agreeing with Weinberg, we know that we should call "complication" what he calls "complexity": as we saw in Sects. 1.1 and 1.3.1, causal complexity does not coincide with mere multiplicity of causes.

The parable of the Cell-Centred Perspective in cancer research shows how ignoring complex causality leads to the complication of causes, at the epistemic level, by multiplying the number and types of causal factors that must be taken into account to explain biological processes. As the number of causal attributions increases, so does the risk that at some point we lose sight of what is actually being explained. Moreover, we don't have the criteria to establish which elements of the system and their interactions, together with external factors, *acquire* causal power (which I will call 'specificity', see Chaps. 5 and 6, especially Section 6.4). A working hypothesis we can put forward following the evolution of the explanatory models of cancer and the biology of cancer outlined in the first chapter, is that causal complexity actually implies the coexistence of different kinds of causal dependencies, while multiplicity of causes simply refers to many causal factors responsible for cancer.

In the next Chapters we will see many more things. The mechanistic-reductionist strategy, by ignoring the epistemological and ontological status of a biological system (a tissue, an organism) able to meaningfully integrate signals, seems condemned to fail. But eventually we will try to understand in which sense and why a Cell-Centred Perspective works, because in some sense it undeniably does, as the research tradition presented in this chapter has shown.

Chapter 3
The Systemic Approach to Cancer: Models and Epistemology

3.1 Overview

In Chap. 2 we have seen that attempts to find an explanation of carcinogenesis at the 'lowest possible level', either in terms of genes or in more general molecular-mechanistic terms, not only hit a wall of complication (Sect. 2.10), they also end up by moving towards the tissue level (Sect. 2.9) without actually abandoning atomistic perspectives and mechanisms. Suitable formal and mathematical systems were available from engineering, where the mechanisms of auto-regulation and complex circuits are defined as a "collection of interrelated elements coherently organized with one and the same end [...] that can manifest a dynamic behaviour, adaptive, seeking an auto-protective objective and evolutionary" (Meadows et al. 1992, p. 105). Systems Biology – a powerful example of a systemic approach in the life sciences – imported these models to observe complex realities with auto-organization characteristics. What is more, during the carcinogenetic process most of the molecular components remain unchanged while their functional activity changes, due to external and internal factors.

Whereas mechanistic models considered functions as incorporated in the parts intrinsically, systemic approaches and Systems Biology look at the system as a whole and focus on the functions that emerge. Among the cultural reasons for the emergence of systemic views and theories, we can count the progressive erosion of traditional reductionist[1] paradigms that are unable to grasp the dynamic regulative properties of complex systems.[2] Systems theory, departing from reductionist and

---

[1] See Appendix for some background on reductionism, and Chap. 2 for examples. Then, see Chaps. 5 and especially 6 for my interpretation of reductionism in the context of a Dynamic and Relational View of cancer.

[2] However, a systemic thought is much more antique in science. In chemistry and thermodynamics such notion of system was already clear and commonly assumed in scientific practice since the end of the nineteenth century. The same can be said about electromagnetic field and electric circuits. In 1952, Tellegen proved a theorem that states the dependence of systems' behaviours on two inde-

M. Bertolaso, *Philosophy of Cancer*, History, Philosophy and Theory of the Life Sciences 18, DOI 10.1007/978-94-024-0865-2_3

mechanistic explanatory frameworks, highlights the unity of the connections and functioning of whole systems.[3] Systems theory interprets natural dynamics based on two assumptions: that *the whole is more than the sum of its parts*, in so far as it has properties that are not encountered in the parts themselves, and that *the parts are transformed once the whole has been integrated.*

In the end, I will argue that systemic approaches are interested not only in accounting relationally for how systems work (the system as a 'functioning unity'), but in how their organization comes about (i.e., their 'functional identity'). Epistemological and methodological issues are always related, and issues of hierarchical control have opened new considerations regarding the relevance of distinctive levels of biological organization. I largely postpone the philosophical discussions of these topics to Chaps. 5 and 6, where I present my Dynamic and Relational View for cancer research. In this chapter, more specifically, I present some systemic models that demonstrate the relevance of interactions, emphasizing synthesis over analysis, and study the cancer phenomenon at levels other than the cellular-molecular one.

3.2 Systems Biology

Systems Biology started with efforts to explain "how the molecules in action determine the characteristics of living systems" (Boogerd et al. 2007, p. 9). The main experimental challenge of Systems Biology consists in integrating numerous and varied data, obtained with the application of new technologies at molecular levels, in a unitary, quantitative and predictive theoretical framework. For some authors, the natural culmination of this work would be a 'virtual cell' (Sect. 3.3.1). This virtual cell, despite being presented as "not merely a more refined illustration but one that offers a new level of mechanistic comprehension" (Editorial, Nature 2006), can be as well seen as a mark identifying Systems Biology as a simple evolution of the traditional models adopted to describe biological processes (Sect. 2.9).

---

pendent components: constitutive laws of elements and emergent laws from their relational structure (Tellegen 1952; Penfield et al. 1970). Tellegen's theorem – and other similar discoveries – have important practical consequences, which are related to the relevance of the relational structure that is irrespective of the single elements that compose the system. I have to thank Alessandro Giuliani for clarifying this and other points, eventually related to what I have called "non-trivial determinism" in the volume (see Chap. 6).

[3]Various authors have confronted this subject since the early decades of the twentieth century. Amongst them, we can mention Goldstein, Jonas, Bergson, Simondon, Ruyer, Haldane, Whitehead. Some philosophers, such as Quine or Duhem, developed extreme conclusions in the philosophical sphere. Some form of extreme anti-reductionism is presented as 'vitalism'. 'Holistic' positions can be found in contemporary biology too. Radical view points have been expressed, for example, by biologist Hans Driesch. In general, the systemic perspectives that we consider here are those inspired by the works of L. von Bertalanffy (von Bertalanffy 1968) or other authors like C.H. Waddington and J. Needham or P. Weiss. We will see how a systemic approach can be compatible with reductionist accounts.

In fact, new forms of systemic analysis became necessary as the possible explanations of traditional models broadened. The classical form in which biological systems are described (be they metabolic charts, gene expression regulation pathways, protein-protein interaction maps, food webs and so forth) is a set of basic elements (genes, proteins, metabolites and so forth) and their connections, and some rules of the kind, 'is transformed into' or, 'is increased (decreased) by'. Groups of basic elements are linked through lines and arrows (Alberts et al. 2002). With the development of high throughput methodologies, these graphs became larger and larger and required some form of global analysis in order to overcome their wild multiplicity. Systems Biology thus ventured not only into explaining, "how the molecules in action determine the characteristics of living systems" (Boogerd, cit.), but also into clarifying how biological processes, regulated at distinct levels of biological organization, *maintain* their *functional unity*. Systems Biology, in sum, deals with living organisms as organized molecular systems *and* as organizers of molecular systems. Indeed, Systems Biology's main interest is perhaps understanding the manner with which new functional properties arise at new levels of biological complexity.

The majority of investigators are aware that the questions raised by Systems Biology go further than simple methodological ones. What I defend is that epistemological and methodological issues are always related, with no possibility to establish any priority for the role of one aspect with respect to another.

From the epistemological point of view, the idea that living beings have functional properties, which cannot be understood by molecular biology, has led to a paradigm change in scientific investigation. In mechanistic models functions are usually incorporated in the parts intrinsically, and they are the *explanans* of identified events and mechanistic effects (e.g., mutation and proliferation). In the new paradigm, the functional properties of the parts are an *explanandum*. Certain reflections in relation to this point have been proposed and assembled in a study regarding the philosophical foundations of Systems Biology (Boogerd et al. 2007), already cited above.

A double origin of Systems Biology has been pointed out, therefore, like in the case of systemic theories. In the words of Krohs and Callebaut (2007), Systems Biology has a double root: a *component root* and a *system root*. The root that looks at the component, centres its attention on the individual macromolecules. When molecular biology used its instruments to develop bioinformatic systems and x-omics, i.e., the large scale studies of molecules and their products such as proteins, through highly efficient techniques, the component root was its natural culmination. On the other hand, the root that looks at the system as a whole is focused on the functions that emerge when multiple molecules interact simultaneously. Krohs and Callebaut have also noted that the two roots might converge. The functional dynamics of a system cannot be explained in terms of interconnected parts but requires a broader perspective directly centred on the questions of control and the organizational regulation of the system itself. Microarray techniques for gene expression analysis are examples of how new techniques have been used to facilitate this approach. The systemic root is represented by control theories and by mathematical models that use non-linear systems, according to the logic of collective phenomena, which are intrinsically stochastic. In this domain, the problems generated by hierarchical control have opened new considerations regarding the relevance of distinctive levels of biological organization. There are functional properties of the cells that cannot be identified on genetic or epigenetic levels, but require higher levels; at least the level of control processes run by micro-RNAs (molecules that do not code proteins, but regulate the genes that produce them) in order to distinguish the functional state of cells, for example, that of hematopoietic differentiation (cf. Felli et al. 2010). As such, this approach sets aside the level of genes in so far as they are molecular parts that act as sources of biological information necessary for the development of the dynamics of a superior organization.

3.3 Systemic Models in Cancer Research

Some spurious experimental findings in cancer research that have encouraged several groups to undertake the systemic route to modelling are the following:

- *Experimental evidence that non-genetic mechanisms can be the base* for the neoplastic process in all of its phases, including the reversion of the neoplastic phenotype, both *in vivo* and *in vitro* (Bizzarri and Cucina 2007).
- *The importance that the whole tissue and organ contexts play* in the initiation and progression of cancer. Normal cells, placed in intimate contact with an inappropriate tissue, degenerate into a tumour, while neoplastic cells inserted into an embryo, or treated with embryonal cytokines, regain many of their normal characteristics (Biskind and Biskind 1944). This evidence highlights that the normal phenotype can be recovered, indicating the stabilizing presence of a strong normal morphogenetic field that gives the appropriate signals for integration and tissue coordination to its neighbours.
- *The intrinsic non-linear dynamic component of endogenous molecular and cellular networks* that react to environmental changes of the cells, modifying their activities by regulating them through homeostatic mechanisms that have evolved in metazoans over hundreds of millions of years (Greaves 2001).
- *The stochastic nature of gene expression and cellular differentiation* (Laforge et al. 2005, see Sect. 3.3.1).
- *The functional heterogeneity of tumour cells*. As previously mentioned, common morphological, immunological or biochemical changes of tumour cells have rarely been described as phenotypic alterations, and no new metabolic pathways have been identified.

A functional property, understood as a dynamic behaviour or functional state at the cell-population or tissue level, is therefore what these studies began to focus on. As we shall see now, the system under inquiry is outlined through concepts that are free from mechanistic accounts,[4] such as morphogenetic field, functional states, functional landscape, which describe functional-structural properties of a higher-level system. The common focus of systemic models is on the causal relevance of the *interactions* among molecular parts, rather than on some causal molecular part. The large-scale network of interactions carries the most important causal relevance in these models. For this reason, systemic models can be seen as falling outside a reductionist-mechanistic approach (Sects. 2.9, 2.10, and Appendix), being more interested in the ongoing dynamics of the biological process under inquiry rather than on the outcome of localized mechanistic interactions. The systemic approach privileges synthesis over analysis. With the term *synthesis* I refer to the process by which a data series is taken together with – and in virtue of – its specific relationships, allowing for considerations of the phenomenon at different levels.

I will now overview some systemic models that I consider particularly significant from an epistemological point of view, for the concepts they use or for the way they frame their own explanatory enterprise.

3.3.1 Modelling the Endogenous Molecular-Cellular Network

In Chap. 2 we have seen how the development of the Cell-Centred Perspective – from the Somatic Mutation Theory to the consideration of more and more factors and mechanisms – culminated in the "emerging integrated circuit" of the cell (Hanahan and Weinberg 2000), an ideally complete circuit of molecules within and across cell boundaries that should provide a full explanation of the pathogenesis of cancer.

Molecular systemic models of the cell connect cellular agents such as ONGs and TSGs, growth factors, and cytokines into an *endogenous molecular-cellular network* with non-linear dynamic interactions. A complex network is a representation of the entities interacting in a system, by means of a graph, i.e., a collection of nodes connected by links (respectively called "vertices" and "edges"). Nodes can take on different states that depend on the neighbouring nodes in different ways. Links can have different directions, weights, and signs; they can be directed (represent unidirectional flows between nodes) or non-directed (showing mutual interactions). The non-trivial topological features of biological networks generate non-linear and stochastic dynamic with different stable local states – phenotypes – that have more or less obvious biological functions. Some of these states may be considered "normal", where "normality" of the achieved states depends on organic conditions (Ao et al. 2008). Cell growth may be "abnormal" if it induces a high immune response or a high-energy consumption, typical of neoplastic masses.

[4] However, mechanistic accounts are, sometimes, implicitly assumed. The concept of functional state, for example, is implicitly assumed in the definition of some explanatory frameworks (like the concept of stem cell, see Sect. 2.7) but not without epistemological implications, as we will see more specifically in Chap. 7.

Transitions from one stable network state to another may be produced by the stochastic character of the system's behaviour. When an endogenous network is not optimized in the interest of the whole organism, the latter may be considered "sick". Properties of endogenous molecular-cellular networks can also be genetically compromised, for example by mutations or epigenetic changes. At times, the alteration may come about as a result of abnormal interactions between tissues or their parts, and special functions may activate when the cells are subject to stress situations. In fact, despite the cellular network being "endogenous", its factors are produced through physical or chemical mediation, making it sensitive to critical thresholds of cellular balance and interaction between cells.

The systemic analysis concerns the number of possible stable states within a "functional landscape" of the endogenous network. The state space is the total set of possible network states, with the transitions and trajectories between them. To highlight some characteristics, the state space can usefully be represented (or thought of) as a tridimensional surface showing the disposition, functionality, and accessibility of the different states (Fusco et al. 2014; Serrelli 2015). Cancer is then seen as a robust state of the endogenous cellular network that is not (yet) optimized for the benefit of the whole organism. Under this assumption, mathematical models can study neoplastic transformation as a changing behaviour of pre-tumour, phenotypically normal cells. In turn, a quantitative description and interpretation of certain aspects of tumour genesis and tumour progression becomes evidence supporting the hypothetical importance of the stochastic component in choosing cell fate (Ao 2007; Ao et al. 2008).

According to the Autostabilization-Selection Model (Laforge et al. 2005), tissue coordination is based on two processes: *phenotypic self-stabilization*, through which differentiated cells stabilize their phenotypic identity, and *interdependence for the proliferation* through which differentiated cells stimulate the proliferation of different phenotypes. Simulations based on the model and on experimental data[5] show that the system is capable of producing tissue organization (Fig. 3.1). Conversely, modifications within the cell that change the balance between self-stabilization and cell interdependence lead to tissue disorganization and affect the control of cellular activities, including gene expression. Cancer, therefore, derives not only from mutations in tumour cells, but rather from a gradual accumulation of many small alterations in the *balance between* self-stabilization and interdependence for proliferation.

---

[5]The model incorporates experimental data from gene expression studies, more specifically about the modulation of the concentration of transcriptional regulators in the cell in relation to differentiation. The model incorporates stochastic gene expression with computer simulation models. It also takes into account similarities with models pertaining to the theories of morphogenesis (cf. with Sect. 1.2.1). One important difference is that in the Autostabilization-Selection Model the molecules only act as stabilizers of a prior state reached stochastically, not as promoters of a change in cellular state, as in morphogenesis. Embryogenesis is the evolution of the first cell, the zygote, as it moves towards the balance mentioned here; instead, cancer is the destruction of the same balance.

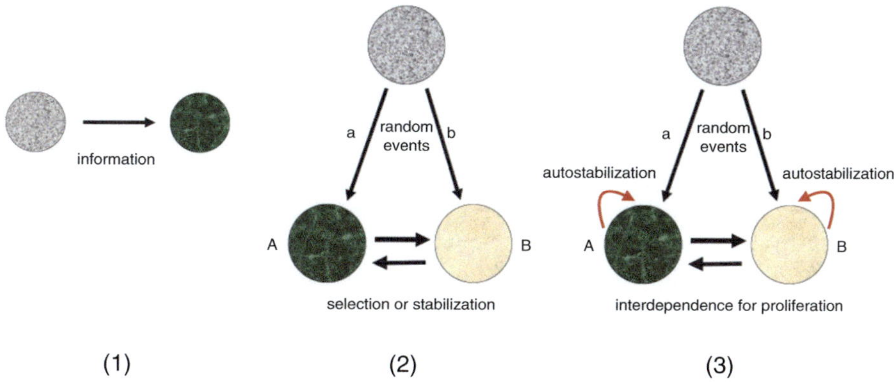

Fig. 3.1 Genesis of the Autostabilization-Selection Model. "(1) Instructive (or determinist) model of cell differentiation. (2) Selective (or Darwinian) model of cell differentiation. According to whether the random event a or b occurs, the cell differentiates into type A or B. (3) Autostabilization-selection model of cell differentiation. Cell differentiation and tissue organization result from stochastic gene expression, interdependence for proliferation and autostabilization of cell phenotypes" (Modified from Laforge et al. 2005)

The interdependence seen in the Autostabilization-Selection Model can be considered a new and relevant explanatory dimension of systemic models. Tissue formation is not the result of a single type of cell interaction but involves multiple and functionally complementary molecules that exert various effects on the tissue, as well as the determinant effect of cellular interactions. Tissue organization and cellular interactions cannot be based merely on selection between various cells (Sect. 2.8). A balance between parameters controls cellular organization and growth. An imbalance can lead to tissue disorganization and invasive growth, like that shown in cancer. In particular, tumour growth is a local effect of an *imbalance* between all the factors involved in organizing the tissue.

3.3.2 Attractor Landscapes and Switch-Like Transitions

Let us now take a step back and talk about the evolutionary assumptions that justify and inform some of the Cell-Centred systemic models.

The SMT and other related theories (see Sect. 2.6) assume that somatic evolution drives the multi-step process that produces a metastatic cancer, but it is difficult to reconcile random mutations with the inexorable progression to metastasis, in virtually all cancers, and the change in phenotype associated with tumour cells, characterized by the Epithelial-Mesenchymal Transition (see Sect. 2.8). In addition, due to the irreversible nature of genetic mutation, it is also difficult to explain how certain metastases remain dormant and can form links with the normal surrounding tissues.

Moreover, an evolutionary imperative for all metazoans seems to be the suppression of mutant cells that would escape their normal limits and move towards independent clonal expansion. Various types of mechanisms are indeed in place to remove the cells that have undergone a process of abnormal cell division. Some of these are intracellular, such as those assigned to the control of cell cycle progression, while others are made of intercellular signals that constrain the cell to remain within the microenvironment that supports it. Together, these tumour-suppressor mechanisms are extremely effective, explaining why cancer occurs less than once in a lifetime, on average, despite the trillions of potentially tumourigenic cells, each the bearer of hundreds of genes potentially responsible for cancer and theoretically subject to a significant number of mutations. Furthermore, our antitumour defence system can discriminate neoplastic cells (by abnormal growth) from normal ones and effectively keep the former under control without suppressing the latter.

A conceptual framework based on the system-level dynamics of gene regulatory networks, can help reconcile these behavioural and evolutionary inconsistencies (Ingber 2008; Huang and Ingber 2000).

For any biological network – e.g., a gene expression network – the existing regulatory interactions between the various genes or molecules determine the *state space*, i.e., the whole set of states that are available to the network and the possible transitions and trajectories between them. A close relationship exists between the particular architecture of the network and the way in which the network's movement through the state space is constrained. Most state spaces feature one or more *attractors*, i.e., balanced trajectories or equilibrium points that, once reached by the network, remain stable also in face of perturbations. On a landscape representation of the state space, a pattern of lines converges to a point or a closed trajectory, that functions as an attractor of all the trajectories originating from a nearby state. What this means is that a state S functions as an attractor when, if pushed to a state S*, the system usually returns to S. This model allows a unification of genetic determinism with an alternative view that emphasizes the importance of non-genetic components on cellular mechanisms and dynamics of abnormal cell bonds in tissue architecture (Huang and Ingber 2006), integrating and expanding to new dimensions the linear view of the mechanisms and biological processes. A genetic mutation can be related to cancer through the rearrangement it causes in the connections of the regulatory network. In other words, a mutation can be related to cancer through the distortion of the state space landscape it causes, turning cancer into a quite stable attractor state.

The idea that the cell's fate or, more specifically, the fate of the various cell types of the body is differentiated by attractors is consistent with a series of biological findings that are difficult to explain by linear causal dynamics, represented by signalling cascades or by genetic pathways. Coherent changes in cellular phenotype, underlying the neoplastic progression to the metastatic phenotype, may result from dynamic (switch-like) transitions within entire genome-wide gene regulatory networks.

One of the fundamental properties of the attractor landscape is multi-stability, i.e., the ability of the system to go back and forth, moving to specific and stable phenotypes in response to a range of non-specific disturbances, including genetic "noise". A reversible switch is then possible, and becomes plausible in the attractor model, as the neoplastic phenotype can return to the normal one.

Now, it can be postulated that embryonic attractors remain present in adulthood, although hardly accessible to cells of the organism in this stage of life. We cannot exclude that these can act as tumour attractors in cases of malignancy development. Thus, ontogeny provides oncogenesis with a starting point (Ingber 2008; Huang and Ingber 2006), and this is due to the self-organizing nature of the programs of gene expression. The existence of "cancer attractors" would suggest that the development of tumours is a matter of regulation of gene expression and selection of a stable, pre-existing program, as is the maturation of the cell type and its differentiation during development (Ingber 2000). The epigenetic character typical of cancer cells, highlighted by many researchers, would be consistent with this hypothesis and based on theories already developed by Waddington at the beginning of the last century (Waddington 1935). It would therefore be appropriate to speak of cancer as a problem of "reprogramming".

Current research on mesenchyme focuses on the role and expression of mesenchyme-specific genes during development and pathological processes, and the locations and capabilities of mesenchymal stem cells (see 2.8). In the attractor landscape perspective, the mesenchyme[6] phenotype should be considered a distinct cellular program, more consistent than the sum of the effects of individual genes, which separately encode particular characteristics. In fact, the molecules that can induce a mesenchymal phenotype in transformed cells vary. This fact suggests that the malignant transformation itself causes a change in the behaviour of the regulatory networks rather than a change in the mechanism involved. Mutations can change other molecular factors that contribute to the distortion of the landscape and to the shift of the system towards a strange attractor (tumour) (Huang and Ingber 2006; Fogarty et al. 2005). In any case, the malignant phenotype is not a complete cellular reinvention, but rather one of the states potentially existing in the cell.

If pre-existing attractors explain the ease with which random mutations can quickly produce a wide range of embryonic features, Darwinian selection, both in terms of attractors and of individual genes, may also gain importance in the progression by modulating the proliferation and optimizing cellular survival. Indeed, the increase in cell number is a key factor because a mutation stabilizes and becomes advantageous within a population of individuals, a phenomenon usually defined as *clonal interference* (Aranda 2002). Somatic evolution is therefore seen as a co-evolution between cells and their microenvironment. Factors that induce a change in

[6] Mesenchyme is a type of tissue composed of loose cells embedded in the extracellular matrix (a mesh of proteins and fluid) which allows mesenchymal cells to migrate easily and play a crucial role in the origin and development of morphological structures during early development (especially those concerning connective tissues, from bones and cartilage to the lymphatic and circulatory systems). The interactions between mesenchyme and another tissue component, epithelium, help to form nearly every organ in the body.

the geometric shape of cells are involved in the change of cell fate, influencing the evolution towards apoptosis, quiescence, proliferation or differentiation.

The creative power of somatic evolution is, however, also limited by the attractor landscape perspective. Many, if not all, networks that drive cell proliferation are intrinsically equipped with growth suppressive properties: they inhibit or eliminate any immediate selective advantage that mutations in these pathways may otherwise give to the cells. Every pre-cancerous cell that acquires a single mutation in an ONG is effectively *trapped within an evolutionary cul-de-sac* because no particular pathway confers a net selective advantage: the intrinsic suppressive activity for growth within each pathway is controlled by another pathway, adjusting the proliferative potential of cells. This systemic balance of the various components is shown also in Laforge's 'autostabilization-selection model' (see above).

The prospect of the *attractor landscape* emphasizes the non-genetic origin of tumour and metastatic phenotype. The functional *cul-de-sac* of cells causes the coordination of extracellular activation of a multiplicity of pathways simultaneously. Interpreting cancer as a problem of epithelial-mesenchymal interactions that become pathological by a breakdown of the basic rules that govern tissue organization leads to the idea that tissue level autonomous mechanisms dominate, and are mediated by, cell-cell interactions rather than by gene-gene relationships.

3.3.3 Regulation from the Cell's Microenvironment: The Dynamic Reciprocating Model and the Extra-Cellular Matrix (ECM)

There still is an open question underlying the models that look at cancer in terms of functional states. This question regards how functional and positional information is interpreted and encoded by the primary sequence of genes and how the environment affects gene expression to form an organism with a multitude of tissues and functions. The issue acquired a special interest when progress in molecular biology revealed the presence of intracellular cytoskeleton complexes that, in addition to having an obvious role in the shape of the cell and its movement, seem invested with an important regulatory function at the genetic level (Bissell et al. 1982).

The Dynamic Reciprocating Model (DRM) postulated that somehow the cytoskeleton, together with the nuclear matrix, is involved in a dynamic reciprocity of inter-relation with the Extra-Cellular Matrix (ECM). There are two subtypes of the ECM: the Interstitial Stroma (IS) and the Basement Membrane (BM), a specialized form of ECM. The IS matrix surrounds the cells with connective tissue, while the BM is present at the baso-lateral surface of different types of cells in many tissues. The BM is composed primarily of laminins, type IV collagen and proteoglycans such as heparan sulphate, deposited by many different types of cells (Kalluri 2003). The basal epithelial cells adhere closely to the BM, for example in the mammary glands, and not only provide mechanical support – while separating the epithelial cells from the stroma – but also determine cell polarity, proliferation, differentiation and gene expression of individual cells (Hagios et al. 1998; Liu et al. 2004; Bissell et al. 2005).

The cytoskeleton constitutes a signal axis between the micro-habitat and the genome that sustains a "dynamic reciprocity" (Xu et al. 2009a) between cells and the ECM. Eukaryotic cells contain three different types of cytoskeleton: microfilaments, intermediate filaments, and microtubules. Cytoskeletal proteins form a macromolecular complex in the points of adhesion between the cell and the ECM by means of signal adaptors and modulators. The involvement of integrins, mediated by the ECM, induces the reorganization of both the active and the intermediate filaments. Laminins, structural proteins that fortify the nuclear envelope, are connected to the actin filaments of the cytoskeleton by means of nesprin proteins that solder the nucleus to the cytoskeleton in order to regulate localization, movement and other functions (Berrier and Yamada 2007; Hetzer et al. 2005; Zhang et al. 2001). On the other hand, cell mechanisms also have a profound influence on the micro-habitat of the ECM, by means of the regulation of gene expression and assemblage of ECM, by means of the action of integrins on the collective ordering of fibronectin and the transcription of Metallo-Matrix Proteases (MMPs, see Xu et al. 2009a).

According to the DRM, the stability of the different cell states depends less on the differentiated cell, and more on the cell plus its ECM (which may or may not be derived from the cell). This model can be described as "the minimum required unit for expression of tissue-specific functions", and is an extension of existing models for how it interprets the membrane-cytoskeleton interaction (Bissell et al. 1982). It also emphasizes that the most typical aspects of these models are not easy to depict graphically: the component of dynamic reciprocity between the cells and their matrix evolves continuously during various types of biological processes.

Is the cell shape to be considered a cause or an effect? In most studies, it is difficult to separate the effect of the ECM and its components from the changes induced in the cell shape. In some cases, like the functional modulation of the mammal glands *in vitro*, the *in vivo* collagen does not clearly carry out the same function of the one used *in vitro*. Four factors could influence the ultrastructural differentiation of the epithelial mammalian cell: the access to nutrients of the basolateral surface, the cellular proximity to the surface, the interaction of epithelial cells with the elements of the stroma and the flexibility of the substrate, allowing for the shape of the cell to change (Bissell et al. 1982). The ECM, as supramolecular organization of connecting molecules, would probably serve as multilevel ligands that can influence the clustering of growth factor receptors. This characteristic would explain the capability of the ECM to simulate the action of growth factors. The clustering *per se*, beyond the interaction with growth factors, has been shown to be able to induce cellular growth (Kahn et al. 1978; Schlessinger 1980; Bissell et al. 1982).

After its introduction, the DRM was extended to provide an integrated view of development, cancer and ageing (see Sect. 1.1.2), asserting that genes are only like "the keys on a piano: although they are essential, it is the context that makes the music" (Nelson and Bissell 2006, p. 287). It is the tissue architecture that is critical for cellular homeostasis and tissue specific functions. The involvement of ECM receptors in the induction of both physical and chemical signalling cascades, which are transmitted from the cell membrane to the nucleus, is accompanied by changes in morphology and cellular and tissue architecture. These alterations involve reorganization of the cytoskeleton and of the chromatin, leading to changes in the cellular and tissue architecture as well as gene expression, all of which, in turn, influence the microenvironment. This dynamic and reciprocal dialogue between cells and their microenvironment acts as a circuit that includes an axis of signal transmission that propagates through, and is regulated by, the cytoskeleton (Xu et al. 2009a).

Cancer occurs then when the dynamic and reciprocal interaction, mediated by the cytoskeleton, is compromised and damaged for an extended period of time. The accumulation of ECM components and the expression of ECM remodelling enzymes are subject to strict temporal and spatial regulation, reflecting the importance of a fine tuned ECM microenvironment in stabilizing the architecture necessary to carry out the functions of a specific tissue. Enzymes, for example, charged with the remodelling of the ECM, like MMPs, are able to modulate the tissue architecture in a normal process of organogenesis, but it has also been shown that forced expression of these enzymes destroys the tissue microenvironment, leading to tumourigenesis *in vivo* (Radisky et al. 2005; Sternlicht et al. 1999; McCawley et al. 2008). This alters the delicate balance between the ECM and cell homeostasis, destroying tissue architecture, and is sufficient, over a long period, not only to impair the normal function, but also to induce tumourigenesis. We can assume that the cytoskeleton is therefore essential in both cellular and ECM regulation, although much work remains to be done to define the mechanisms involved (Xu et al. 2009a).

Nevertheless, there are experimental data showing that, in laminin-rich Extra-Cellular Matrix (lrECM) 3-dimensional (3D) cultures, the expression of membrane receptors, located primarily in the areas delegated to the cell-cell junctions, is dramatically lower in non-malignant cells, while remaining unchanged in malignant tumour cells. Therefore, how a non-malignant cell communicates with its 3D context, both at the cell surface as well as at the nucleus, differs from that of a malignant cell. The ability to convert to the malignant phenotype by activating these membrane receptors and protein signals shows that the acquisition of a malignant phenotype is accompanied by changes in the architecture and tissue by reversible changes in protein expression that allows a transformed cell to circumvent the strictly hierarchical events inherent in the normal differentiation process (Spencer et al. 2007).

3.3.4 Systemic Approaches to Genomics and the Causal Role of Genomic Heterogeneity in Cancer

As we know, high heterogeneity of mutations and patterns of genomic alterations has been described in cancer (see H2 in Sect. 1.4.2):

> Examples include the fact that most of the karyotypes of solid tumors are drastically altered compared with the normal human karyotype (…); there is a significant correlation between

karyotype heterogeneity and poor prognosis; and the recent finding that some regions of the genome are conserved by organismal evolution but altered in cancers (…). In addition, there are many sub-types of the same cancer, and it is possible that the same tumor can evolve from multiple cell lineages. It has been proven that even a single cell can generate cells with drastically different karyotypes as this stochastic process generates heterogeneity (Heng et al. 2009, p. 540).

Genome sequencing of huge numbers of human tumours was carried out in an attempt to identify common sets of genes (Collins and Barker 2007) for a unified description of the neoplastic phenotype, at least in genetic terms, but the results, as anticipated by some authors, were disappointing (Greenman et al. 2007): despite the initial success in the identification of a large number of mutations with high penetrance in certain populations of patients, the great part of these mutations have low frequencies among patients in general; also, the continuous increase in the number of genes identified as responsible for various forms of cancer makes it doubtful that one can ever arrive at a series of mutations actually common to all types of cancer.

Without somehow abandoning a cellular vision of cancer, attempts were made to analyse, in quantitative and dynamic terms, the degree of heterogeneity at the genomic and epigenetic level (Ye et al. 2007). Individual molecular pathways have limited predictability during the stochastic evolution of the tumour, where genomic heterogeneity, reflected in karyotypic changes in cells, dominates. Unlike studies centred on the analysis of specific genes, a type of models (e.g., Heng et al. 2009) considers the pattern of evolution *of the system* (i.e., the change of its functional state), rather than the specific pathways compromised in the neoplastic process.

The assumptions of this model are: (1) that tumour progression is an evolutionary process where the genomic reorganization of the system, rather than of a single pathway, constitutes the driving force (Heng et al. 2006a, b, c); (2) that the potentially unlimited number of genetic and epigenetic alterations that take place during the evolution of cancer makes it practically impossible to identify a common mechanism at such level; (3) that heterogeneity is therefore to be considered a key feature of cancer and not a background noise to be eliminated in the epistemological approach to a complex phenomena like tumourigenesis. Far from being an independent confounding variable,[7] *heterogeneity* is actually an intrinsic feature, conferring complexity, and strictly linked to the system's *robustness*. The "true challenge […] to understand the system behavior (stability or instability)" (Heng et al. 2009, p. 539) can be met through an adequate understanding of the role that genomic heterogeneity at systemic level plays in the evolution of cancer:

Heavily influenced by reductionism's view, most of the molecular analyses of cancer have focused on a molecule of interest, without considering the overall status of the genome system. It has been generally assumed that during molecular manipulation or specific targeting the bio-system remains the same. This assumption has been pushed to the extreme where genome level information has become largely ignored by most of the molecular

[7] Some discussions pointed out how classical physical science attempts to reduce noise by simply increasing the sample size. However, in complex biological systems this does not solve the issue of heterogeneity, because variability is not simply a "noise" tied to a specific experimental approach (see also Heng et al. 2008).

> analyses. The fact is, however, when the overall karyotype changes, the role of the same gene may also be altered, as the function of genes are dependent on their genetic network, which is defined by the genome context (Heng et al. 2009, p. 540).

In cancer, systems continually change during progression and this is illustrated by significantly altered karyotypes as well as expression patterns (Klein 2002; Ye et al. 2007; Heng et al. 2008). "From a system point of view, significant karyotypic changes represent a 'point of no return' in system evolution, even though certain gene mutations and most likely epigenetic changes can influence karyotypic changes" (Heng et al. 2009). The prognosis will then be linked to the stabilization of determinant karyotypes in tumour cells.

The models considered here do not take cancer as a progressive, step by step development, but as a phenomenon of stochastic (macro)evolution.[8] The stochastic events referred to here are not completely random. Rather, they introduce a level of unpredictability because of the differences in *initial conditions* that can be deduced from the genetic and epigenetic alterations detected at different levels of genomic organization. Therefore, as in macroevolution, tumour progression is considered to be fundamentally different from the process of development. The latter refers to well-regulated processes of self-organization, in both spatial and temporal terms, where many genes play a key role in maintaining the correct sequence of events. In the evolution of cancer, instead, although some cases may present the typical features of a developmental process, most cases are dominated by change: "genome mediated stochastic system replacement, which does not follow a well controlled pattern" (Heng et al. 2009, p. 540). On such grounds, "cancer development" has been labeled an oxymoron.

If one accepts that tumour progression is a sort of macroevolutionary phenomenon mediated by the genome, even research strategies need to change: "Heterogeneity is the reason universal mutations cannot be found" (Heng et al., cit.) and it is also the reason why most of the patients present a panorama of mutations that coincide only in part and have frequencies we cannot consider epidemiologically significant. Note the change in perspective: in the Cell-Centred perspective, tumour heterogeneity was explained by progressive accumulation of mutations; here, heterogeneity is the reason why common mutations cannot even be identified.[9]

Despite the difficulties in establishing a causal relationship among individual molecular mechanisms within complex biological systems, it seems relatively straightforward, for some authors, to establish a causal relationship between *heterogeneity* and the evolution of cancer. Karyotypic heterogeneity is, for these authors,

---

[8] Taking inspiration from evolutionary theory (Serrelli and Gontier 2015), this literature employs the terms "macroevolution" and "microevolution" for demarcating genetic mutations from genomic alterations, punctual events from dynamic processes. In this sense, a macro-evolutionary change would be an overall change of the organizational dynamics, which might give way to cancer (cfr. Heng et al. 2009).

[9] A philosophical frame for this problem is proposed in Sect. 6.4, with the idea that "stability wins over specificity" in biological entities. That is, relational stability of a system is the viability condition for causal specificity of specific events (e.g., genetic mutation) within it.

the driving force of cancer, associated with a large number of genetic mutations and seemingly random epigenetic changes. Only in relatively stable stages (see Sect. 6.4) do DNA damage, karyotypic substitution and gene mutations, as well as epigenetic regulation, play a dominant role, similar to what occurs in the phase of adaptive micro-evolutionary phenomena (Heng et al. 2009). Not so in cancer, where general instability appears to be constant throughout.

As we have seen, the first novelty of a systemic perspective on the genome is that heterogeneity is not a side effect of cancer, but a cause of its final outcome and part of its explanation. Physiologic heterogeneity provides a chance for the success of biological systems to adapt and survive. Dis-integrated heterogeneity creates an environment in which some cell populations can shift their functional state to a neoplastic one. This would be the reason why, according to advocates of this perspective, deriving explanatory principles from simple and homogeneous experimental systems cannot work in the real world, where heterogeneity is not the exception but the rule. As can be deduced from the different karyotypes and profiles of genetic mutations, each cancer appears to be a self-contained example of somatic evolution that, unlike normal physiological processes, does not follow a reproducible pattern of steps. The patterns are defined by the genomic and environmental context (Heng et al. 2009): changing the environment, a specific pattern may become rare as it ceases to be essential to a given process, and vice versa (Hillenmeyer et al. 2008).

3.4 From Systemic Models Towards Systemic Epistemology

Systemic concepts foreground relationships rather than parts. Through a systemic approach, the explanatory factors and the functional properties of cancer are abstracted from their specific biological components. This abstraction allows, for example, a reconciliation of the genetic and epigenetic theories of cancer and focuses on the number and identity of the possible states.

Network (Sect. 3.3.1), *landscape* (Sect. 3.3.2), and *functional state* (Sect. 3.3.3) are different perspectives to identify a system and explain its behaviour. The first perspective focuses on relationships among components. The second one emphasises the sensibility of the system's dynamics with respect to perturbing factors. The third one is more focused on the systems' context and time dependent performance. If the landscape view basically concentrates on response to external 'inputs' (e.g., the motion of the system in an energy field), the functional state has more to do with the concept of phenotype. i.e. with the emerging functional behavior of the system, its 'output'.

Systemic approaches undertake the challenge to analyse the system as a whole, and to study its spatial-temporal development. From an epistemological perspective, this implies acknowledging actual ontologies at different levels of biological complexity. The application of Systems Biology to cancer research thus opens the question about the explanatory relevance of emergent properties in biological sciences (see the Appendix), which are object of debate in both philosophy and science. Taking as default the existence of emergent properties unexplained from an exclusively analytical reductionist perspective directs us to the structural and organizational dynamics of the system.

This, however, does not mean that each level is independent from the lower ones. On the contrary, the regularity and functionality of a level may be dependent in different ways on those at a lower or higher level. Different factors contribute to the maintenance of a functional state, or (like in cancer) to its loss. They act as stabilizing factors rather than causal factors in the mechanistic sense.[10] A cellular system moves to specific and stable phenotypes in response to a range of non-specific disturbances, including genetic 'noise'.

Cancer biology requires a comprehensive explanation accounting for how different levels of complexity interact and operate in a synergistic manner. In this light, two interpretations of the systemic approach are viable (cf. Nicholson 2010; Bechtel and Richardson 2010):

(a) The first interpretation draws the attention on the composition of the system through its elements;
(b) The second interpretation, more radical, but compatible with the first one, tends to see in the system a relational structure where the isolated element disappears, or rather, is defined by the properties that it acquires as part of the system.[11]

What seems clear is that a systemic approach moves far away from a definition of a simple set or aggregation of parts that define the behavioural properties of the system. That being said, interpretation (a) is clearly characterized by a *bottom-up approach*: we have an *organization of the parts, that is, parts have interactions* (usually functionally defined) that take place at the level of the final organization of the system. In this Chapter we have seen systemic explanatory concepts that emphasise the functioning *unity* of systems, beyond the analysis of constitutive parts[12] and mechanic feedback loops. There is constant reference to reflexive and synchronic features of the *dynamics that hold* the neoplastic process and recurrent reference to auto-stabilization and inter-dependencies of the models of tumour cells' behaviour. In (b), the investigated system is viewed from the outset in terms of factors identified within a given organizational level and that functionally belong to it.

Scientists can therefore legitimately state that, "Organisms are clearly much more than the sum of their parts, and the behaviour of complex physiological processes can not be understood simply by knowing how the parts work in isolation" (Strange 2005, p. C968). That is, the dynamics of complex physiological processes *cannot* be understood by knowing how the parts work separately. We need to understand how emergent behaviours, distributed control, and system robustness *are*

---

[10] For a philosophical discussion of this dimension of causality see Sect. 5.2.

[11] There is a wide range of bibliography regarding this subject. I wish to mention: Lewontin and Levins (2007), Urbani Ulivi (2011a), Boogerd et al. (2007).

[12] Contributions from distinct approaches regarding this subject from: Huang and Wikswo (2006), O'Malley and Dupré (2005), Murillo (2010), Hull (1981).

generated (Bruggeman et al. 2002; Westerhoff and Kell 2007). The biology of cancer reveals, beyond the statement that 'the whole is more than the sum of its parts', the question about how 'the parts are actually informed by the whole and therefore depend on it'. We will deepen these epistemological and ontological issues about the *mutual dependence* that characterizes the part-whole relationship in biological systems in discussing the Dynamic and Relational view I am going to propose for cancer research in Chaps. 5 and 6.

Chapter 4
The Tissue Organization Field Theory and the Anti-reductionist Campaign

4.1 Introduction

Cancer researchers, Carlos Sonnenschein and Ana Soto are among the most vocal critics of the Somatic Mutation Theory of cancer (SMT) (Sect. 2.2). The two scientists, leveraging on the features and types of cancer that are not explained by the SMT and its derivatives, propose an alternative theory: the Tissue Organization Field Theory (TOFT), according to which neoplasia arises from a problem of three-dimensional organization of a tissue, not from a normal cell gone awry by mutation or by other mechanisms. The TOFT is based on two considerations (Sonnenschein and Soto 2000): that *proliferation is the default state* of metazoan cells; and that the phenomenon of cancer must be examined from the perspective of the hierarchical organisation of the organism (hence I will speak of an Organism-Centred Perspective). The basic hypothesis is that a tumour is a phenomenon resulting mainly from a defect in the interactions between the cells and other components of the tissue. The organism is a "society of cells" (Sonnenschein and Soto 1999) and cancer is caused by a loss of the proliferative control which is normally maintained by contextual inhibition on the cells. Carcinogenesis does not necessarily require genetic mutations: aberrant stimuli compromising the coordination and structure of the hierarchical organization of cellular systems in metazoans are sufficient to produce cancer.

TOFT supporters, encourage scepticisms and perplexities among some readers and generate philosophical discussions, especially for their appeal to emergent properties and holism. Nevertheless, as a growing literature shows, these authors must be acknowledged the great merit of having opened the door to a deeper review of topics and explanatory issues in both science and philosophy.

M. Bertolaso, *Philosophy of Cancer*, History, Philosophy and Theory of the Life Sciences 18, DOI 10.1007/978-94-024-0865-2_4

4.2 The Tissue Organization Field Theory

According to Sonnenschein and Soto (2000), the Tissue Organization Field Theory (TOFT) is based on two premises, distinguishing it from the SMT:

- The default state of metazoan cells is proliferation[1];
- Cancer must be examined from the hierarchical perspective of the organism and defined as a problem of tissue organization.

The textbook principle that the default state of metazoan cells is quiescence began to be identified and criticized by Sonnenschein and Soto in the late 1970s. Proliferation had long been recognized by microbiologists as the default state of bacterial cells (Soto and Sonnenschein 2006a), and in evolutionary terms it seemed unlikely to Sonnenschein and Soto that single-celled organisms could completely rewire their default state to quiescence while evolving into specialized cells in metazoans. Conversely, animals should have evolved intricate biochemical systems to repress proliferation, not to stimulate it. Soto and Sonnenschein began to formulate their theory in about 1980 with experimental studies seriously questioned the role of the Epidermal Growth Factor (EGF) in promoting the proliferation of endometrial cells. Experiments had demonstrated that the EGF was not a necessary component of the pathway that mediates estrogen action. They had showed that in mice whose estradiol receptors were inactivated there was no proliferative effect on estrogen-target cells; therefore, the stimulatory effect of EGF implants in normal mice did not occur through the estrogen receptors.

An alternative hypothesis was tested experimenting with 'tissue recombinants' of epithelium and stroma interactions. The 'stroma' is, in animals, the supportive structure of an organ, usually composed of connective tissue, and distinct from the 'parenchyma' (the functional parts). Tissue recombinants were sex steroids acting on stroma-target cells inducing them to secrete growth factors that would in turn stimulate the proliferation of the adjacent epithelial cells. The experiments thus showed that the induction of cell proliferation by sex hormones could take place through the interaction between the stroma and the epithelia of estrogen-target cells.

Soto and Sonnenschein – according to their own reconstruction – began doubting the "quiescence theory" and proposing the TOFT as a radical alternative view of cancer. In 1999 they published *The Society of Cells*, arguing that sporadic cancers (i.e., In this context, 'not inherited') arise when pathogens or carcinogens disrupt normal biological interactions between, say, the parenchyma and stroma. To support their view Sonnenschein and Soto collected laboratory and clinical data on spontaneously regressing tumors, a "stubborn fact" that does not fit the Somatic Mutation

[1] The idea that cells do have a "default state" seems to imply a kind of essentialism. Surely different cells, in different contexts and different stages of development, will have a different range of possible states and/or different probability distributions. But "default" here is explanation-relative. That is, it identifies what does not need explanation. The models presented in the previous chapters assume that proliferation needs explanation. Accordingly, those models assume quiescence as the default state.

Theory.[2] Sonnenschein and Soto then went on to contend that the epithelium, at least initially, "can be coaxed back to good behavior" (Longtin 2005).

The 'field' notion in the TOFT refers to the mode in which cells are organized in tissues (Soto and Sonnenschein 2010). It is endowed with causal priority over parts, it has causal relevance in determining carcinogenesis, and it explains tumour heterogeneity. When the structure of tissues is affected cells are "disoriented", and no longer constrained, they cannot differentiate properly. When tumour cells are not integrated in the structural and functional organization of the tissue, they exhibit their original and innate ability to multiply and migrate. The loss of proliferative control is thus to be considered as just one concurrent event.

For the TOFT carcinogenesis is attributable to a process similar to an organogenesis that does not reach completion (Soto et al. 2008a). The organization of cells (the default state of which is proliferation) into tissues and organs in higher organisms is the result of interaction between hierarchically organized cells. This entails the relevance of cell interactions at organismal level: the interactions between cells, mediated by membrane proteins that recognize paracrine, mechanical splices or endocrine signals acting at a distance, are responsible for the transmission of signals leading to the proliferation and differentiation of cells.

Another important implication of the fundamental assumptions of the TOFT is that *cell proliferation is to be considered chronologically removed from the control of the cell cycle*. The latter takes place at subcellular level in the hierarchical organization of metazoans. *Timing* of cancer pertains to a context of affected relationships among coupled biological rhythms[3] and long-range spatial interactions, adding an interesting level of systemic analysis to the overall explanatory account.

The authors of the TOFT connect carcinogenesis mainly to the initial stages of tumour formation. Later stages are seen as the result of a *dynamic process* that can be temporarily or permanently suspended, liable to reconvert to the normal state or to advance towards the stage they call "overt neoplasm" (frank neoplasia; a more specific term than "tumour" rarely used by these authors), or invade and metastasize. The final stage of such a process depends on the *persistence of the same (permissive) conditions that caused the original breakdown* of organization at the tissue level. The malignant tumour phenotype takes on a specific connotation. It is not the necessary functional effect of a causal event. It is better understood as an intrinsic potentiality of cells that in tumours is realized with no adequate contextual control. The *natural history* of cancer is, in sum, told from a new perspective.

Transplantation experiments on recombinant tissue demonstrate that what maintains cell phenotype appropriate to different levels of cellular organization and

[2] As a reviewer of *The Society of Cells* pointed out, epithelial cells mistakenly revert to pro-growth patterns of behaviour by "switch in behavior", not by "mutational meltdown".

[3] Here I prefer the term "rhythms" to "clock", because the clock notion still recalls a mechanical framework where the time dependency is a mere dependency of movements from the relationship between space and time. The concept of rhythms, instead, adds a chronological dimension that characterizes biological processes and that is overlooked by mechanistic-reductionist attempts to explain biological behaviours (see Appendix vs. Chaps. 5 and 6).

tissue in adults are the interactions between epithelium and stroma and parenchyma. In this sense, *tissue architecture* is an emergent property of the cellular society and not a simple function of the collective properties of the cells that constitute it. Although TOFT authors acknowledge that – in some cases – molecules and physical forces do have a causal role, the central issue of cancer and its explanation remains a *problem of the three-dimensional organization of a tissue*, which cannot be reduced to the causal role of its underlying layers. In fact, solid tumors have a distinct structure that mimics that of normal tissues with their two distinct but interdependent compartments stroma and parenchyma. Cells have a memory system in so far as they bear the signs of where they come from (historical information) and where they are (positional information) and it is this information that constrains their future and restricts the differentiation and movement options available to them. The memory of the cell is not necessarily stored in the DNA but can also be found in morphological features.

The TOFT has implications for classifications and terminologies related to explanations in cancer research. For example, the TOFT makes it irrelevant whether the malignant phenotype is considered dominant or recessive (Harris 1986): from an organic perspective it makes no sense to frame the question in such terms. Indeed, what is seriously questioned is the whole research program based on the apparent acquisition of new properties by tumour cells in their progressive transformation to metastatic cells (Shih et al. 1979; Varmus and Weinberg 1992; Land et al. 1983). The "hallmarks of cancer" approach (see Sect. 2.8), focuses on the autonomy of tumour cells from other cells and from their environment, trying to establish a list of the properties of tumour cells. But tumour cells are wildly heterogeneous and not completely autonomous. Relying on these weaknesses of the SMT, the TOFT casts a radical doubt about the explanatory role of cancer cells as such.

There is some ground in common between the TOFT and SMT. Both attribute a crucial role to cell differentiation rather than proliferation. The cells of an organism, when cultured in vitro in a suitable growth medium, seem to survive and multiply indefinitely.[4] As Sonnenschein and Soto write:

> metazoan cells in culture show properties that they did not show at the organismal, nor the organ, nor the tissue level. As they become free from the bonds of homeostatic influence necessary to coordinate the needs of a multicellular organism, the "liberated" metazoan cells, in culture, may reacquire latent ancestral properties including proliferation and motility (Sonnenschein and Soto 1999, p. 80).

The rate of cell multiplication (measured by the doubling time of the cellular population) is not necessarily higher in tumours than in normal tissues. This parameter therefore does not constitute a valid criterion to evaluate or even define tumour cells. For this reason, when talking about tumours, TOFT authors avoid

[4] This concept of "de-emergence", even if only rarely taken up in the literature of the TOFT, is the principal key to understanding the evolutionary dynamics underlying the neoplastic process, a phenomenon that assumes the typical characteristics of an adaptive process in which topological form and organisation of the historical pattern of gene expression are the elements responsible for the failure of the normal program controlling differentiation and morphology.

"proliferation" – that should be considered common to any tissue – and prefer to use the technical terms "hyperplasia" and "hypertrophy" to indicate, respectively, an increase in number of cells and an increase in cell size in the tumour mass (Sonnenschein and Soto 1999). However, although cell differentiation may seem a bridge between the TOFT and SMT, the TOFT frames differentiation in terms of tissue organization and not of properties of the cells. The *hierarchical organization*, considered as the main characteristic of multicellular organisms, primarily entails the *regulatory dynamics* of the organ/organism itself, and the features of cancer cells are to be considered pathological with respect to this assumption.[5] Organism growth (or development) is the right context to find an explanation of cancer (again TOFT authors also avoid using the term "growth" preferring instead "hyperplasia" and "hypertrophy", explained above).

4.2.1 Experimental Approach of the TOFT

The perspective of the TOFT decidedly determines precise experimental approaches (Sonnenschein and Soto 2000), indicating that the theoretical differences "are not inconsequential because the premises favoured by scientists will determine the type of experimental design that they will follow to explore their guiding hypothesis" (Sonnenschein and Soto 2011, p. 657).

First of all, the TOFT fits well with the common, viable diagnostic procedures for cancer. The preliminary diagnosis is usually carried out organically by doctors examining the symptoms and external signs presented by the patient. The pathologist then provides the final diagnosis through microscopic analysis and interpretation of a histological biopsy conducted where they suspect neoplasia. This level of analysis corresponds to the tissue level of biological complexity.

Instead of looking for molecular elements that can be the key in explaining cancer, the TOFT experiments test cellular functioning by acting on tissue dependent features: shape, metabolism and membrane structure. With respect to cells either proliferating or becoming quiescent, the cell cycle mechanisms become virtually irrelevant in light of this theory. The cell cycle is regarded as an automated program – a series of algorithms representative of cause-effect dynamics that are appropriate for mechanical events – so that, from one cell, two are formed.

Rates of cell multiplication change, for example with varying hormone concentration, and this strengthens the argument that cancer cannot be interpreted in terms of cellular autonomy (Sonnenschein and Soto 1999).

Pieces of supporting evidence for the TOFT come from experimental transplantation of tumour tissue (cf. Sect. 1.3.1.) When a tissue is capable of inducing the formation of a new tumour in a host organism, it does so starting from parenchymal

[5] Cell death is not directly discussed either, as it is considered as related to the maintenance of cell number, and not involved as a control mechanism of cell differentiation (Sonnenschein and Soto 1999).

cells. These cells, specific to an organ, are capable of reconstructing the stroma – the supporting structure of the parenchyma – around the new tissue. But first, the pre-existing stroma is lost: it is as if the epithelial cells, altered in some way, are perceived by neighbouring cells as different and these in their turn respond to the environmental changes produced. Eventually there is a weakening of the signals that maintained and controlled the cells that finally express only the phenotype appropriate to the environment of their position: the cells then may start to multiply.

When the components of the tissue cells that form the epithelium and the underlying stroma are artificially separated they stop carrying out the functions performed when they were assembled in their unique original three-dimensional organization. Once recombined, they form a tissue similar to that of their origin (Soto et al. 2008a). When skin cells are placed in a culture they form a uniform layer of tissue different from the original, but if they are placed on a surface previously covered with basic membrane proteins, they tend to group and recover the original three-dimensional structure of the epithelium from which they came. They do, however, often undergo genetic transformation and imbalance when cultured *in vitro* for a long time.

Other pieces of evidence come from tumour regression. Experiments performed using teratocarcinomas and embryonic environments show that the regression of neoplastic phenotype (i.e. the return to normality of tumour cells) can happen (see Sect. 1.2.1). TOFT researchers claim that their theory is able to make sense of such spontaneous regression from a neoplastic phenotype, observed with higher frequency than would be expected were it due to regressive mutation or secondary suppressor mutations (Soto and Sonnenschein 2005).[6]

For the TOFT, a tumour is a phenomenon resulting mainly from defective interactions between cells and the other components of the tissue, and cancer is caused by a loss of the proliferative control maintained by the control mechanism inhibiting the cells. It does not necessarily require genetic mutations. In the crucial phases of cancer onset, aberrant stimuli affecting the coordination and structure of the hierarchical organization of cellular systems in metazoans are sufficient. The effect of carcinogens on structures and subcellular organelles and DNA is thus not directly responsible for the development of cancer: genetic mutations constitute an epiphenomenon that cannot be included among the relevant explanatory causes of cancer (Sonnenschein and Soto 2000); chromosomal and metabolic abnormalities would beconcurrent, not causal, events in the onset of the neoplastic phenotype (Soto and Sonnenschein 2005). Even inherited cancer-related mutations – which generally do not exceed 5 % of cases – owe their causal role to the fact that they already affect

[6]As mentioned in Sect. 1.3, regression is inconsistent with the assumption that cancer depends on dominant genetic mutations and that these are the necessary and sufficient conditions for a reductionist-mechanistic explanation. As seen in Chaps. 1 and 2, the immune system can become detective and destroy precancerous cells, and the strong SMT is only a starting point for the development of a wider Cell-Centred Perspective in which many mechanisms (e.g., epigenetics) concur with genetic mutations and can interact to explain a variety of outcomes. Still, Cell-Centred models – TOFT authors argue – are less able to account for the incidence of tumour regression.

tissue organization in some way. Cancer predisposition is reinterpreted in the TOFT as the existence of tissue-level permissive conditions and the causal role of genetic mutations is mediated by the tissue.

The TOFT admits the concept of "predisposing mutations" but translates them as mutations that create permissive conditions by affecting tissue organization. In other words, TOFT authors see the causality of predisposing mutations as tissue-mediated rather than cell-mediated. Some studies of the APC gene, linked to inheritance of colon cancer, seem to support such interpretation. The protein encoded by APC is, by nature, pleiotropic and it is defined as a "shuttle-protein". Formed by functionally distinct portions, which possess nuclear localization and export sequences from the nucleus, APC is involved not only in regulation of the cell cycle through the Wnt signalling pathway, but also in cell differentiation and the stability of the phenotype through, for example, the maintenance of cytoskeleton integrity, cell-cell adhesion and cell migration (Kemler 1993; Hough et al. 1997; Jaffe 2005; Wasan et al. 1998). In two articles (Soto and Sonnenschein 2004, 2006b) TOFT authors argue that inherited genetic lesions can be relevant as far as they are related to tissue organization and hence SMT crucial empirical data can be reinterpreted within TOFT, as in the case of APC.

To sum up, the TOFT emphasizes that tissue interactions rather than molecular changes make cells normal or aberrant. The mutant genes responsible for hereditary errors in development and cancer find their place in the TOFT (Soto et al. 2004), but the explanatory context of their role is histological and developmental, rather than cellular. The cell is no longer regarded as a virtually independent entity, governed internally by its genes.

4.3 Polarization: Further Evidence Supporting an Organismic Perspective on Carcinogenesis

It is important to remember, at this point, that tissue architecture is relevant also in models that make no reference to the TOFT.

For the TOFT, cell polarity is the first level of biological complexity of a living organism. But cell polarity is an important characterizing feature of tissue organization beyond the TOFT. By definition, polarized cells exhibit an asymmetric internal distribution of proteins, including the membrane receptors and other factors that mediate tissue differentiation and function. The asymmetric protein distribution in polarized cells adds or subtracts inhibitor ligands to their effectors in order to regulate the specific genetic expression of the tissue (Xu et al. 2009a). Some studies give evidence that cells do not express the milk proteins as a response to hormonal treatment in 2D cultures because the basolateral receptors are inaccessible to the apically presented prolactin, and the receptors for prolactin are localized on the basolateral surface of epithelial mammalian cells (Xu et al. 2009b; Ben-Johnathan et al. 1996). These studies, on the whole, suggest that the asymmetric localization of important signal modulators in tissues is crucial for the activation and maintenance of tissue-specific functions.

Other forms of polarity loss can be recognized in different properties of cancer.

The function of an organ relies on the cell constituents *and* on their general organization. It is the obvious uniqueness of this structure that, for example, differentiates a breast from a kidney and that directs the cells of the first to produce milk while the others filter blood in order to produce urine. This is the case even if all cells share the same genome. The structure of an organ is critical for its function and therefore normal architecture can act as a powerful "tumor suppressor", as in some way Harris (2003) already hypothesized, using the term to qualify one collagen form. This form would be capable of preventing the malignant phenotype, even in cells that have been altered by large genetic anomalies (Harris et al. 2007; Mintz and Illmensee 1975; Weaver et al. 1997; Howlett et al. 1995; Wang et al. 2002; Kirshner et al. 2003). The destruction of tissue structure usually goes with the loss of the tissue specific cell differentiation, suggesting that tissue architecture is intimately related to its function (Hagios et al. 1998; Bissell et al. 2003).

If the function of an organ and its homeostasis are guided by the architecture of that same organ and if cells in every organ carry the same genetic information, how are – in such case – specific form and function structured? Some experiments of developmental biologists give very elegant explanations, by postulating that the specific function of the tissue is obtained by means of *interactions* between cells and their context (Bissell et al. 1982).

Among the systemic models presented in Chap. 3, the Dynamic Reciprocating Model (DRM, Sect. 3.3.3) is particularly interesting here. The behaviour of the cell is largely determined by its interactions with the Extra-Cellular Matrix (ECM), with the neighbouring cells and with other causal soluble and systemic factors (Nelson and Bissell 2006). The DRM elegantly shows that tissue specific architecture and function are regulated by the biophysical properties of the ECM, and in particular by the constitution of the matrix, independently, or in addition to, other microhabitat properties.[7] The integrity of the ECM is responsible for stabilizing normal tissue and its functional architecture by dynamic interaction (Xu et al. 2009a), coordinating gene expression and maintaining the tissue homeostasis (Xu et al. 2009b).

ECM molecules can send signals to the nucleus. The ECM acts on cells by anabolic and catabolic processes with mechanical and biochemical signals that ultimately mold the program of gene expression in the cell and affect cellular processes such as activation of transcription factors for survival, or apoptosis. This is done through the activation of the ECM-Response Elements (ECM-RE). In addition, activation of these ECM induced elements involves mechanisms that provoke a change in the chromatin structure and organization (Spencer et al. 2007).

The dynamic exchange between cell nucleus and the ECM is coordinated by an axis of signals channelled through the cytoskeleton (Bissell et al. 1982).

[7] In addition to the biochemical signals that originate from ECM receptors, biophysical and mechanical properties of the microhabitat of the tissue are necessary for cell differentiation, its function and architectural maintenance.

The cytoskeleton notoriously works as a connecting structure between the ECM and the nucleus, allowing for dynamic and reciprocal interactions between the extracellular environment and the nucleus. The reorganization of the dependent cytoskeleton contributes to the transduction of the biochemical signal that stabilizes the maintenance of the architecture and function of the tissue.

In this way, the DRM describes how gene expression is influenced by the microenvironment. The ECM appears to affect both the translational and the post-translational machinery that can influence events through transcription feedback loops. It is increasingly clear that translational regulation plays an important role both in the development process and in the maintenance of tissue-specific functions (Bissell et al. 1982).

The ECM is thus considered an integral determinant of tissue specificity so that, according to the DRM, the functional unit of higher organisms is not the cell by itself, but the cell with its microenvironment (Spencer et al. 2007). The tissue and cellular context play an important, analogous role in development,[8] in differentiation and in the homeostasis of many organisms (Nelson and Bissell 2005).

The ECM is constantly influenced by physiological agents such as growth factors, cytokines and hormones, thus changing over the development arc of an organism, according to the phase of aging or tissue repair phenomena or tumour development. (Spencer et al. 2007). On the other hand, the ECM is strongly influenced by cell activity, causing a change in the composition and structure of the ECM, which in turn influences the cell and so on. "Tissue architecture is then both a consequence and a cause, the end and the beginning" (Nelson and Bissell 2005). This kind of circular interaction is the keystone of systemic approaches (see Chap. 3) and will be further elaborated in Chap. 5 with particular reference to the asymmetry of the causal relations between cells and tissue (see Sect. 5.2).

4.4 The TOFT and an Organism-Centred Anti-reductionism Campaign

The TOFT emphasizes the importance of (micro) environmental factors and in many aspects plays down the autonomy of cells, moving towards an account of the neoplastic phenomenon at higher levels of biological complexity. In this way, the TOFT can be viewed as proposing an organism-centred perspective that is clearly different from the Cell-Centred Perspective described here in the early chapters.

In the next sections it will be seen that the TOFT yields important epistemological implications. For example, it entails acknowledging that real emergent properties exist. Cellular components, once integrated into tissue, interact in a way that presents new and unique functional features. Complex processes, such as

[8] At least for the organs of placental mammals, organs that develop similar structures do so by utilizing the same environmental factors, also because ECM responsive elements (ECM-RE) are evolutionarily conserved, at least functionally (if not in their nucleotide sequence).

carcinogenesis, bring into play the hierarchical control of a biological organization, and heterogeneity is better understood.

Advocates of the TOFT have the great merit of opening the door to deeper study of these explanatory issues both in science and philosophy as shown by an extensive literature (Marcum 2005; Malaterre 2007, 2011; Bizzarri et al. 2008; Bertolaso 2009a, b, 2012a, b; Plutynski 2013). However, it will be argued, these practical and theoretical proposals need further elaboration to grasp the deep epistemological meaning of the multilevel phenomenology of cancer. The philosophical discussion set off by the TOFT appears sometimes confusing, especially when the theory is proposed in the context of a campaign against "reductionism". The rhetorical efforts to clarify the scientific presuppositions of the TOFT leads to simplify the theoretical framework and to ultimately miss the causal relevance of different levels of biological organization.

4.4.1 The Holistic Version of TOFT

The authors of TOFT define their theory as "holistic" and "organicist", suggesting a sharp contrast between the TOFT and the SMT, and conveying the idea that explanations of cancer must be sought in *either* parts *or* wholes.[9] The TOFT that emerges in this debate is arguably a particular version of the theory. But this is not the only way of presenting the TOFT: there is also a systemic version of the theory. Before examining the systemic version of the TOFT, the holistic one will be examined with the consequent TOFT-SMT debate and the resulting issues of incommensurability and incompatibility between the two theories.

Holism acknowledges higher-level effects as causally relevant to the maintenance of the functional structure of a system. According to Scott Gilbert and Sahotra Sarkar, in a paper frequently cited by the TOFT advocates, the holistic perspective claims that "when we try to explain how the whole system behaves, we have to talk about the context [...] on the whole and cannot get away talking only about the parts" and that "complex wholes are inherently greater than the sum of their parts in the sense that the properties of each part are dependent upon the context of the part within the whole in which they operate" (Gilbert and Sarkar 2000). The holism that characterizes the TOFT depends on the postulate that cancer is an emerging phenomenon occurring at tissue level. The study of cancer and its development cannot

[9]A possible philosophical analysis of SMT and TOFT could conclude that they are looking for "principles of order" – responsible for the specificity of living beings – in different things, *either* parts *or* wholes, analogously to the historical debate between mechanism and vitalism (Wolfe 2012). For a discussion, see Bertolaso (2013a), where I argue that the question of greater interest is not so much to evaluate at what level the biological organization encounters a "principle of order" as it is to demonstrate how this is manifested. I have reframed the question concerning a "principle of order" in terms of "principles of organization", complexifying the "modes of action of causality" in the structuring of living organisms (see Sect. 5.1).

thus be reduced to a complex series of interactions between proteins (Sonnenschein and Soto 2000).

Consistently, holistic TOFT proposes the following epistemological framework: (1) emerging properties exist; (2) the whole is greater than the sum of the parts as far as these emergent properties cannot be explained only in terms of properties directly attributable to the individual pieces; (3) this implies that to understand a specific biological phenomenon each hierarchical level must be studied without expecting that lower levels will necessarily contribute to our understanding (4) a phenomenon cannot be studied independently of the level at which it is observed. *Therefore* (5) top-down causality is the most adequate assumption to explain complex mechanisms; (6) normal tissue architecture acquires a significant importance, while mutations are just epiphenomena.

Implications (5) and (6) are peculiar to holistic TOFT, and, although introduced by 'therefore', do not in fact logically follow assumptions (1–4), as a simple logical analysis reveals. TOFT supporters, appealing to emergent properties and to a holistic position to make sense of their scientific model, drive some readers into scepticism and perplexity.

While the holism referred to stresses the relevance of the whole with respect to the parts, a different claim is "organicism" – also called "materialistic holism" in the literature (Gilbert and Sarkar 2000). Organicism "has provided the philosophical underpinnings for embryology since the time of Kant" in the eighteenth century; it claims that "wholes are so related to their parts that not only does the existence of the whole depend on the orderly cooperation and interdependence of its parts, but the whole exercises a measure of determinative control over its parts" (Ritter and Bailey 1928, cit. in Saetzler et al. 2011). Implicit in this description is the concept of *emergence* – "the idea that at each level of biological organization new properties become manifest, that could not have been predicted from the analysis of lower levels" (Saetzler et al. cit.) – along with that of *mutual dependence* we have mentioned by the end of Chapter 3.

While rejecting any vitalist position by being 'materialistic', TOFT's organicism adds to the claim that it is impossible to explain higher-level properties in an organism from a mere reductionist perspective: the organism as a whole would influence and determine in the first place the properties of its parts.

Holistic TOFT seems to entail a claim about *the most relevant* level of analysis in causal terms. But the TOFT in its systemic version, explained below, only requires an ontological commitment to *the existence of* emergent properties. These two issues are not homogeneous in kind and should be addressed separately. They are nevertheless often mixed up in the context of attacks against reductionism, as in the following quote:

> A prevalent philosophical stance in biology is methodological reductionism, which predicates the study of biological systems at the *lowest possible level* with the objective of uncovering molecular and biochemical causes. In this view, causes act from the bottom-up. Contrary to reductionism, organicism considers *both bottom-up* and *top-down causation* (Soto and Sonnenschein 2010, 2, my emphasis).

Beyond emphasizing the real existence of *emergent properties* (e.g., tissue properties), i.e. higher-level properties that cannot be predicted from the properties of parts, the TOFT defense argues that *downward causation* is *the most appropriate* prerequisite in research on complex phenomena such as cancer.

There are venues in which the authors of the TOFT relate the causal relevance of emergent properties to a *systemic view* (Bunge 2003, cited in Sonnenschein and Soto 2005) rather than to a holistic perspective of the biological phenomenon. The systemic TOFT is more explicit about the fact that the operation of composing and decomposing in understanding biological dynamics and behaviours, especially when embedded in inter-level regulatory processes, is not trivial nor are the two operations mere opposites. The systemic TOFT aims to *combine* the synthetic approach with the analytical one for understanding emergent phenomena, overcoming the reductionism-anti-reductionism dichotomy. In that systemic view bottom-up causal relationships (i.e., causal interactions among parts of the system yielding system-wide effects) coexist with top-down causality exerted by the relationships that make these interactions causally relevant. Indeed, the two directions of causality imply each other. Unfortunately, as we will now see, the TOFT is often presented in the context of anti-reductionist battles, where its more organicist and holistic versions are emphasised.

4.4.2 Polemic Targets of TOFT

There are many ways to characterize TOFT's polemic targets. The genetic account of cancer originally supported by the SMT seems to be a paradigmatic example of all such targets, although the precise critical points are often conflated or confused in the debate between the TOFT authors and their adversaries. Since cancer research is very varied as we have seen in the previous chapters, different mixes of these polemic targets can be spotted in the debate about the explanation of cancer.

The first polemic target of the TOFT is *reductionism*, and the TOFT is labeled an "anti-reductionist" theory. The issue of the lowest level of explanation matches the traditional philosophical discussion about epistemological and ontological reductionism in biological sciences, which state that a privileged level of explanation should eventually be found at the physical level and its laws. We may say that anti-reductionism focuses on the specificity of systems and on the dynamism that characterize the proper object of biology. Biology deals with phenotypes and behaviours (or, more synthetically, *behavioral phenotypes*). Anti-reductionism defends behavioral phenotypes as 'new' properties that are really existent and causally relevant in nature. Scientists talk about emergent properties, while philosophers often frame the discussion in terms of supervenience, bracketing the real existence of emergent properties and how such existence should be understood (see Chap. 5 and Appendix).

Advocates of the TOFT also claim to be reacting against the *mechanistic* account supported by the SMT. The SMT's aims are identified with the explanatory objective presented in the famous "hallmarks of cancer" paper (Hanahan and Weinberg 2000): "notwithstanding that genes and proteins known to be involved in the neoplastic process are now counted in the hundreds, evidence suggests that the occurrence of all cancers is governed by a common set of mechanisms that are limited in number" (Hanahan and Weinberg, cit.).

For the TOFT, the SMT's commitment to mechanistic causal accounts of cancer's dynamics is eventually closed in a sort of *genetic fundamentalism*. Although, as the evolution of the explanatory models show, a mere genetic explanation of carcinogenesis necessarily overcame the original simplification, the possibility to eventually explain the origin of cancer as a mere result of molecular alteration events ends up in a circular, embodied system of mechanisms that is far from being explanatory of anything. SMT statements of the sort that: "a change in the DNA of a somatic cell alters its characteristics so that it undergoes clonal expansion" (Vaux 2011, p. 341), underpinning the idea that "it is much easier for a single mutation to transform a cell that is already dividing, or already has stem cell characteristic" (*ibidem*, 342), are clearly open to criticisms and challenged by the mentioned data regarding reversibility of the neoplastic phenotype of tumour cells.

TOFT authors also criticize SMT's atomism, i.e. the idea that there are "fundamental components" that eventually explain everything (cf. also Sonnenschein and Soto 2013a). If mutations are the *fundamental* causes of cancer, cancer is to be considered *as an autonomous cellular process*. The tumour cells are normal cells that have gone "crazy" (Weinberg 2006). And, strictly speaking, we should assume that the SMT's proponents are not using the image of a 'renegade cell' as a metaphor for the *explanandum*, but also of the explanatory framework, i.e. as a constitutive element of the atomistic assumption. The SMT assumes that the cell is the *fundamental unit* of the organism (Sonnenschein and Soto 2011, p. 338). Of course, atomism can be either a methodological claim (we can investigate the biological organization of the organism *in terms of* cells) or an ontological one (organisms are nothing but aggregations[10] of cells). We can look at the organism as an aggregation of cells (structural dimension) or as a network of cells (functional dimension) depending on the aspect we are interested in or able to investigate. But the position of the SMT on this point is very ambiguous and clearly open to the TOFT's criticism: "According to the somatic mutation theory (SMT), cancer begins with a genetic change in a single cell that passes it on to its progeny, thereby generating a clone of malignant cells" (Vaux 2011, p. 341).

TOFT advocates appeal to the claim that no genetic program exists to account for development and thus carcinogenesis. Besides reinforcing the discussion on the relationship between the epistemological and ontological dimensions of reductionism in biological sciences, this criticism also strikes *biological determinism*. This latter point is interestingly summarized in the following quote (Soto and Sonnenschein 2010):

---

[10] We adopt this term of aggregation because it maintains the same meaning across different fields and authors (Bechtel and Richardson 2010; Wimsatt 2007; Dupré 2010; Silberstein 2008).

> Starting in the 1960s, metaphors such as that genes were in the 'driver's seat' and the introduction of the term 'developmental program' persuaded generations of researchers, teachers and students that development was the mere unfolding of a program encrypted in our genes.[11]

Gene-cell centred views have been widely contested by developmentalist and organicist supporters as (1) the sequencing of several genomes has revealed that the number of genes in a given genome does not correlate with the complexity of the corresponding organism and that these gene numbers are too low to trigger development from DNA to phenotype; (2) no univocal correspondence occurs between a DNA 'gene', the several RNAs produced from it by splicing, and the resulting proteins. But neither reductionism nor genetic determinism could operate without a clear one-to-one correspondence between gene and protein.[12] Experimental biologists who embraced an organicist approach criticized the claims of *reductionists* and *genetic determinists*, and proposed new, dynamic and integrative approaches.

The TOFT also accuses the SMT of being *untestable*, on the basis of empirical data[13] and of research procedure (see the argument about the functional test in Sect. 6.4.1). TOFT's supporters, in fact, say that the SMT "is – with current technologies – essentially untestable" because it is currently really difficult to separate and manipulate a single DNA chain. However, subordinating the testability of the SMT to the improvement of biotechnologies just postpones the problem, and does not

---

[11] Confusion could here emerge again from quotes like this. In concrete, if we are not comparing two theories but a theory with a metaphor, the problem might just consist in how far and why these metaphors work. If, instead, we are talking about criticisms of the SMT's position things might be different. I will take the former stance as TOFT privileged point of view. In any case, I have to acknowledge that the natural reaction of philosophers could be to say that scientists make use of philosophical claims when they are in trouble and cannot further explain their position within their own conceptual field. Whether this might be the case for the TOFT is, in my opinion, an open issue. Surely, their style in contrasting the SMT has not always been of help in clarifying the epistemological issues at stake and to get the most out of their own reflections and methodological approach. Moreover, reading this and other quotes reported above, might make seasick any philosopher for which each term, from metaphors to emergence, reduction and multiple-realizability, top-down and bottom-up causation evoke an almost infinite literature on quite different topics and issues. Clearly, some issues do not seem to be adequately framed. Nevertheless, TOFT did spur many to reflect and consider some paradoxical issues that everyone working in molecular pathology is confronted with practically.

[12] The TOFT authors define this problem as the problem "of the many and the many". In this definition, they cite philosopher David Hull. However, Hull used to refer to this problem as "one to many" (Hull 1976). The latter usage is much more represented in philosophy.

[13] I report them in footnote for clarity but also to avoid redundancy in the text with what is already presented earlier in the chapter and in previous chapters. Following *Bioessays* papers the main points to claim against the SMT are (cf. Sonnenschein and Soto 2011, p. 337): (a) normalization of tumour cells remain unexplained in the SMT, such as in the case of spontaneous regression of neuroblastoma; (b) regression of hormone-sensitive tumours and their metastases; (c) *normalization* by regulation of *tissue architecture*; (d) neoplastic induction by ectopic tissue transplantation; (e) neoplastic induction of epithelial cells by altered stroma.

present any argument against an in-principle testability of the SMT.[14] The epidemiological data regarding the inherited tumour risk, related with some mutated genes, is not entirely conclusive about the genes' relevance in cancer origin and onset.

More criticisms to the SMT arise on the basis of unexplained empirical data and inconsistent results, as TOFT authors put it. The SMT would "fail to provide an explanation for observable phenomena in cancer biology" (Sonnenschein and Soto 2011, p. 338). However, this could mean either that the SMT is completely wrong or that the TOFT explains more about cancer than the SMT (more on this topic will be said below).

The TOFT can be seen as part of a broader perspective, just like SMT was revealed as the seed of a Cell-Centred Perspective: the TOFT may be seen as the battering ram of a non-reductionist organism-centred perspective characterized by the explanatory relevance of emergent properties. The Cell-Centred Perspective endorsed by the SMT is clearly inadequate if we have to take it as a statement that cancer can be explained, without any loose ends, just considering the autonomous behaviour of a cell fully determined by its genome.

4.4.3 The Epistemological Limits of the TOFT

SMT authors entered the discussion only recently, trying to defend their position against the TOFT. As we have seen, although recombinant DNA techniques in the 1970s led to enormous progress in identifying genes involved in the phenotype of tumour cells, it soon appeared that we are still a long way from being able to identify precisely the genes responsible for cancer. The claim that "Sooner or later, the process of cancer pathogenesis (disease development) need[s] to be explained and understood in molecular terms" (Weinberg 2006, p. 54) could be understood not as a commitment to genetic determinism but, eventually, only to genetic reductionism, where genetic molecules might have a privileged explanatory status. That is, if one were interested in genes, the problem would not be to find the genes responsible for the origin and final onset of cancer, but to present an explanatory model in molecular terms, where such molecular parts are genes. Consider the following statements taken from SMT-based papers. Exceptions to the reversibility of the neoplastic phenotype like the case of embryonic stem cells and teratocarcinoma cells (that are considered in any case rare by the SMT but are an argument against the SMT for the TOFT), can be explained by the SMT in terms of epigenetic changes (cf. Vaux 2011, p. 342). Moreover, "[c]ell clonality is essential to SMT, cell autonomy is not" (Vaux 2011, p. 342). It is not excluded that such an evolution of the SMT arguments could be just due to the discussion that the TOFT raised, social and cultural factors

---

[14] Note that this point is nevertheless relevant and worthwhile to be investigated. Experiment design and control that are routinely performed in research papers and programs are some of the issues related to this point.

being considered for the progress of science. However, what is of interest here is that these statements seem to provide evidence that the SMT has been changing, forced by a non-reductionist dimension in the definition of the *explananda,* so that the real issue of the reductionist – anti-reductionist debate in cancer research is not genetic determinism but became what a reduction is in biological sciences and what place mechanisms have in it (see Chap. 6 and also the Appendix).

Beyond the multiplicity of causes in the etiology of cancer that are still the focus of different research programs and epidemiological studies, the two co-existing explanatory theories, SMT and TOFT, paradigmatically represent the general tension that emerges from the *search for key mechanisms* in the account of the neoplastic process. What characterizes the debate is an almost obsessive search for key parts or mechanisms at the subcellular level that can account for the particular behaviour of tumour cells. The TOFT seems to be raising 'simply' the problem of justifying the adoption of a determined level of hierarchical organization for the explanation of a phenomenon such as neoplastic formation.

In my opinion, one of the points that is not well addressed in the TOFT and explains part of the scepticism is that it does not convincingly show why, once all the relevant units and their interactions are identified, prediction could not follow for higher-level properties as well. Moreover, what also remains an open issue is why molecular alterations cannot be considered in such interactional terms, as some research programs seem to assume in the TOFT based studies. While stating that tissue organization is the only explanatory relevant level, in practice the experiments of the TOFT, too, proceed gradually with exploring the lower levels of organization and organic complexity.

Another criticism that the TOFT faces is the risk of indulging in an attitude that might appear as equally reductionist when moving *from the claim re the relevance* of microenvironment and tissue architecture for the origin and onset of cancer, *to the defense of the tissue level as the only* proper level of inquiry, *and then to the assertion* that from this perspective genes do not matter at all and mutations in cancer are just epiphenomena. As the TOFT also asserts, in fact, cancer is a disease of the hierarchical organization of the organism. So genes should matter, and the problem is to clarify how and why or in which sense. This is a crucial point in the debate.

In the view that I shall propose in the next chapters, I will point out that both these problems are typical of a concept of emergence that still uses a part-whole ("mereological") language. Within the mereological framework – adopted by both the TOFT and the SMT through the part-whole talk – an explanatory model that adopts a top-down causation still looks for mechanisms responsible for some major feature of the behaviour of the system at a specified level of analysis so there is no substantial difference between the two accounts of SMT and TOFT. Malaterre (2007, 2011) made this point. The proposal will be to abandon the primacy of any part-whole framework and claim that a *theory of fields* in biological sciences is the most important aspect of the epistemological proposal of the TOFT. The dynamism of biological systems will be shown as more fundamental than the resulting (and variable) part-whole organization, and will also highlight the importance of choosing the explanatory level in scientific practice, which does not seem to be self-evident or merely determined by pragmatic interests. These issues will be crucial to understand the reductionist – anti-reductionist debate in cancer research. But for now the following is a brief address on the issue of causality as it manifests in the TOFT-SMT debate.

The issue of causality is well articulated due to the peculiar relationship that holds among causes and effects in biological behaviours. TOFT supporters say that in science there often is a confusion between causes and explanations (Sonnenschein and Soto 2008). In particular, the SMT "[c]onflates the notion of *causation* with that of *explanation*" (Sonnenschein and Soto 2011, p. 658, my emphasis). However, framing the issue in these terms, the whole picture is complicated. The SMT says that in fact "[t]he relationships between DNA mutation, cell proliferation and the development of cancer are complex and circuitous, so it can sometimes be difficult *to separate cause and effect*" (Vaux 2011, p. 342, my emphasis). The TOFT, from its own perspective, says that "there are many interactions that occur simultaneously to maintain the structure of a tissue; hence, it is practically impossible to sort out cause and effect in a way that would precisely reveal whether emergents have true causal agency" (Soto and Sonnenschein 2005, p. 115). The difficulty is differently solved by the two theories. The SMT limits itself to acknowledging a multiplicity of causal factors organized in a molecular circuitry. The TOFT assumes that a top-down causation is at work. Therefore researchers can either "take for granted that emergent phenomena exist and adopt an organicist stance, or alternatively assume a reductionist stance hoping that a neat, linear causal chain will eventually be identified" (*ibidem*). But, while claiming that top-down causation is different in kind, i.e. plays a different role, from bottom-up causation, the TOFT is not able to explain in which sense.

Issues about top-down and bottom-up causality are not well framed within a discussion of emergence as such, as the TOFT tries to do. As noted by Malaterre, the question that immediately follows is of course: which kind of emergence? "Various concepts of emergence have been developed, weaker versions being compatible with a reductionist view while stronger versions not" (Malaterre 2007, 9). Following the analysis of Stephan (1999), Malaterre explains that three of the major positions (weak emergence, synchronic emergence and diachronic structure emergence) posit physical monism (i.e. entities classified as emergent are instantiated by systems consisting solely of physical parts) as well as synchronic determination (i.e. the fact that there can be no difference in the systemic properties without there being differences in the properties of the parts or their arrangements).

Differently from Malaterre's analysis,[15] which takes into account only TOFT's arguments, I think that the appeal to emergence entails more radical features of cancer and biological behaviours. Higher-level properties have an explanatory priority in any process that implies inter-level regulatory mechanisms. A *synchronic*

[15] I clearly largely agree with Malaterre's analysis, although I aim to expand and deepen the reflection here and especially in the next chapters. One point of agreement with Malaterre which I have not cited yet is his argument that evidence should be found within the multilevel phenomenology of cancer, rather than in its analogies with other morphogenetic processes.

reflexive causation is at work and needs to be taken into account by the explanatory models: the cause-effect relationship between tissue and cells – and between any whole and its parts – is asymmetric and synchronic, and cannot be reduced to any diachronic effect either.

The different causal dimensions will be dealt with in Sect. 5.2. Although downward mechanisms are perfectly compatible with bottom-up causation and its related mechanisms, it will be argued that a synchronic reflexive downward causation (cf. also Kim 1999) does require a different analysis. Cancer biology is paradigmatic of this causal dynamics,[16] as the analysis of tumour heterogeneity shows. Such causation has ontological implications not because it claims for vitalistic forces in the causal account of biological processes, but because the inter-level regulatory dynamics of cancer development can be better understood by acknowledging a different framework to account for part-whole relationships than the mereological one (Sects. 5.2 and 5.4). Taking into account such complexity of causal dependencies is first and foremost an epistemological result of scientific practice in the identification of the explanatory system (Sect. 6.3), and it is not sufficiently emphasized in usual defenses of the TOFT.

4.5 Incommensurability and Incompatibility Between TOFT and SMT

Although TOFT's authors often refer to incommensurability as a synonymous of incompatibility (e.g. Soto and Sonnenschein 2005), I think that a philosophical distinction between the two is needed. Then it will be apparent that the claim of the TOFT about the incommensurability of the two theories do not play in favour of the TOFT itself.

Are TOFT and SMT incommensurable? Some arguments have been presented to argue that they hardly can be. As Malaterre (2007) pointed out, strictly speaking, incommensurability would mean that their terms and *objects of inquiry* could not be translated into one another. The two theories would be incommensurable if

> …one sticks to a very strict and restrictive formulation of each theory, that is, if we take the SMT as a theory looking solely for genetic causes of cancer and the TOFT as a theory investigating solely tissue organization causes of cancer, then the two theories may well appear incommensurable for they would appeal to mutually incompatible explanations and causal factors (Malaterre 2007, p. 12).

Since both SMT and TOFT seem to converge towards the same levels of biological organization (tissue elements) and both need to confront genetic elements in their arguments, it is very unlikely that they can be declared incommensurable (see Malaterre 2007, 13). Their scientific question is the same (carcinogenesis) and their

[16] Kind of causation of whose coherent sense Kim rightly doubts from the point of view of the formal structure and logic of scientific explanation (reported by Malaterre 2007, p. 12). Some arguments against Kim's scepticism are presented in Mitchell (2012) too.

explanatory hypotheses aim to make sense of the same empirical observations, such as aberrant proliferation of cells, tumour cells heterogeneity, epigenetic changes, heavy genetic alterations, etc.. Incommensurability, in this case, would be intrinsically contradictory (on this theoretical and philosophical issue, an interesting study has been carried out by Buzzoni 1995).

If the two theories are not incommensurable, then the question is what explanatory framework could make sense of their convergences.

Are TOFT and SMT, instead, incompatible? Consider, for example, their assumptions regarding the default state of cells in metazoans (quiescence for SMT, proliferation for TOFT). Soto and Sonnenschein write:

> …while the somatic mutation theory (SMT) of carcinogenesis assumes that the default state of cells in metazoans is *quiescence*, the tissue organization field theory (TOFT) of carcinogenesis posits that, to the contrary and consistent with evolutionary theory, *proliferation* is the default state of all cells. These significantly different outlooks make the two theories incompatible*; that is, that they are lodged at different levels of biological organization* [my emphasis] and, therefore, address phenomena that occur in different places: namely, inside a cell for the SMT and in tissues for the TOFT (Sonnenschein and Soto 2011, p. 657).

This quotation – like others that can be found in TOFT commentaries – conflates three points of divergence of the two theories: (a) holding different assumptions about the default state of cells, (b) being incompatible, and (c) focusing on different levels of organization. Focusing on the discrepancy between assumptions (a), we might accept that the two theories are incompatible. But if then we think that the differing assumptions direct the two theories on different levels of biological organization (c), then the differing assumptions do not demonstrate incompatibility. In the TOFT, the assumption that the background state is proliferation is indeed tied to the adoption of a holistic perspective. For the SMT proliferation just belongs to the methodological framework that looks at cancer from a cellular point of view, and makes use of functional and active terms to describe cells behaviour. Proliferation is not a main explanatory issue, but an implicit assumption when the focus is on the behaviour of a tumour cell that starts proliferating in an aberrant way.

The same kind of confusion is found in quotes concerning the function of metaphors in science:

> Metaphors and images have been used in order to shed light on the subject of explaining cancer. The SMT centres on "one renegade cell," and views cancer as a cell-based disease involving unregulated cell proliferation. The TOFT, instead, focuses on a "society of cells" and views cancer as a problem of tissue organization. Hence, as hinted above, explanations of the process of carcinogenesis by these two theories belong to distinct levels of biological complexity and, therefore, are *incompatible*, as are their philosophical stances (reductionism versus organicism […]). The above-referred incompatibilities do not rule out, however, that the data gathered from experiments based on the SMT might be interpreted either in the context of the TOFT, or even to refute the arguments of the SMT (Sonnenschein and Soto 2008, my emphasis).

Again, if the SMT and TOFT metaphors are metaphors in the same sense, i.e. if they entail an explanatory dimension of the *explananda*, the two theories might in fact be incompatible; but if, as the TOFT authors seem to say, such metaphors frame the *explanans*, then we might be dealing with just different explanations of the same phenomenon, and that doesn't imply incompatibility. Evidently it could be said that metaphors such as the 'renegade cell' would work within certain explanatory ranges. For that matter, it could be pointed out that the difference between the two theories is methodological, and that the TOFT supports systemic approaches based on its different assumptions than molecular-mechanistic models.

I am not satisfied with a discussion of the two positions as having different *explananda* and different methodologies because, first of all, there is a claim of inadequacy of reductionist explanations, which is of great philosophical (more than scientific) interest. Secondly, and consequently, what actually matters is *how* two different explanatory accounts can make sense of the same phenomenology and finally seem to converge towards the same level of explanation. Indeed, the most intriguing question is whether and why, i.e. to what extent, the two explanatory theories could be considered compatible. It will be argued that, although TOFT authors give a negative answer to the question "Would it be productive to reconcile the SMT and the TOFT?" (cf. Sonnenschein and Soto 2008), what is at stake is a deeper understanding of how *they might imply each other*. What we need to explore is whether the disentangling point is to show how different models black box different causes providing independent explanations, or rather to show that they can be considered epistemologically tied together, i.e. that one model cannot be confirmed without the other.

4.5.1 A Peculiar Kind of Compatibility: SMT as a Specific Case of TOFT

A way out of the debate, that could be more comprehensive than arguing for a radical replacement of SMT with TOFT, could be to understand clearly the epistemological relationship between SMT and TOFT claims. Setting aside, for the moment, the conflict about the real *explanans* of carcinogenesis, the TOFT has much to say about the *explanandum*: that is, the TOFT *defines the system* to be studied. Given this, the best argument of the TOFT against the SMT revolves around the relevance of the context. Scientific accounts of carcinogenesis do not simply imply questions about the mechanisms involved in cancer progression but also about the epistemic conditions for those mechanisms to be explanatory. In fact, the context is what confers an *explanatory power*, although limited, to any SMT-related model. Yet, the explanations of the SMT avoid the issue of context dependence.

We have seen that cancer shows a sort of natural history in which cells and their context dynamically interact over time. Tumour latency and cell heterogeneity are different manifestations of that (Sects. 1.3 and 1.4). Tumour cells are not completely autonomous, as acknowledged by both the SMT and the TOFT. Tumour cells have 'relational' properties that do not pertain to cells as such (e.g. epigenetic modifications and dependence on the microenvironment integrate the final accounts) and they are 'functional' in a very specific way. Evidently, statements about the incomplete autonomy of the behaviour of tumour cells imply acknowledging the relevance of the biological microhabitat in the origin and development of cancer (see above, in particular Sect. 4.3). On this point, TOFT authors are quite convincing: "[C]ellular context is critical to carcinogenesis, a concept about which the fundamental [in the sense of original] form of the SMT is silent" (Sonnenschein and Soto 2011, p. 337).

The SMT does not have the epistemological tools to take this explanatory dimension into account. It simply overlooks the issue. Consistently with its atomistic view of the cell behaviour, explanatory emphasis is on how cells do what they do: "in a tissue with proliferating cells, oncogenic mutations will occur more frequently, and (…) mutations can both increase the rate of cell proliferation and cause genetic instability leading to faster growth and yet more mutations" (Vaux 2011, p. 342).

> Science is systematic organization of knowledge about the universe on the basis of explanatory hypotheses which are genuinely testable. Science advances by developing gradually more comprehensive theories; that is, by formulating theories of greater generality which can account for observational statements and hypothesis which appear as prima facie unrelated (Ayala 1974, xi).

This statement has been taken very seriously by TOFT authors[17] trying to make sense of the difficulties in explaining cancer and its dynamic properties. The TOFT keeps a firm focus on the wider context where cancer develops (the tissue, the organism etc.). That does not mean that the conceptualization of the context is well-defined and consistent. As we have seen above, big difficulties arise in sorting out *causes and effects*, coming not only from the huge number of relationships and interactions, but also, importantly, from the reciprocity of causal effects, that the circularity characterizing complex biological behaviours entails. In fact, the context is a complex biological system. TOFT authors explicitly refer to such circular aspect when considering the reciprocal causal influence of the biological structures of the basement membrane and its epithelium: "How can causation be studied here? Is the tissue causing the formation of a basement membrane? And then, is the basement membrane causing the normal architecture of the epithelium- and then the tissue? This looks like circular causation", (Soto and Sonnenschein 2005, p. 115). A satisfactory understanding of this apparent circular causal influence involves not only experimental science but also epistemological reflection, as different levels of investigation overlap and each experimental approach requires the individuation of a level of analysis that reduces the phenomenon to the typical dynamics of the chosen level (cf. the discussion about the *mesoscopic level* in Sect. 6.3). It is evident that the reciprocity of the dynamic interaction in complex biological phenomena makes it

[17]TOFT authors quoted Ayala and a very similar citation in a recent article on "Why Systems Biology and Cancer?" (Sonnenschein and Soto 2011): "…(1) "science seeks to organize knowledge in a systematic way, endeavouring to discover patterns of relationship among phenomena and processes; (2) (s)science strives to provide explanations for the occurrence of events;" and finally, "(3) (s)science proposes explanatory hypotheses that must be testable, that is, accessible to the possibility of rejection" or falsification" (Ayala 1968).

difficult to determine all efficient causes and cause-effect relationships, as required by the experimental method. "Acknowledging these problems does not seem an exercise in nihilism, but a first step towards trying to devise ways of studying organisms while taking into consideration the problems posed by their historicity" (Soto and Sonnenschein 2005, p. 106).

The circularity of effects acknowledged by both the SMT and the TOFT does not leave a safe way out to the question about cancer causes. But the SMT does not merely omit higher-level causal dependencies. It subsumes these causes within the assumptions of its Cell-Centred models. For this reason, the TOFT is epistemologically more powerful and comprehensive than the SMT and can be generative of other explanatory accounts different from the tissue one.

Heritable cancers, in which a genetic account seems inescapable, represent the strongest point of contact between the SMT and the TOFT. Although the TOFT prefers to distinguish these different kinds of cancer; at least in two articles (Soto and Sonnenschein 2004, 2006b) TOFT authors argue that inherited genetic lesions can be explanatorily relevant as far as they are related to tissue organization, so that the crucial empirical data of SMT can be reinterpreted within TOFT, as in the case of APC (Sect. 4.2.1). In hereditary tumours, the role of the context and of the organizational inter-relations is implicit in the proper functional definition and identity of the genes. Under these conditions of validity, the notion of gene can have an epistemological privileged status, although not a metaphysical one.[18] Therefore, the SMT can be – to some extent, i.e., as far as it works like in these cases – considered a specific case of the TOFT when it is brought into a unitary epistemological perspective, which is systemic. sporadic cancers – more appropriately explained by TOFT – and heritable cancers, in which a genetic account seems inescapable. The SMT was able to identify real causes, as in the case of the Rb gene and retinoblastoma (Sect. 2.3, footnote) or the APC (Sect. 4.2.1). Acknowledging these successes implies conferring to the argument a real explanatory value, even if a limited one. But the relevance of APC or p53 in cancer risk can be explained in the TOFT as a tissue-mediated cause. Inherited tumours and risk can be more satisfactorily explained in the TOFT as well (Sect. 3.4).

[18] Soto and Sonnenschein criticize the still dominant assumption of a metaphysical primacy of genes: "[A] great number of biologists insist that explanations should always be sought for at the gene and/or gene product level, regardless of the level of organization at which the phenomenon of interest is observed. This stance, genetic reductionism, together with its twin, genetic determinism, predicates that everything in biology may be reduced to genes because the genome is the only repository of transmissible information" (Soto and Sonnenschein 2005, p. 104). A famous advocate of such position is Richard Dawkins. "In [his] view, genes are the only units of selection (Dawkins 1976) and *development is just the unfolding of a genetic program*. In sum, genes in this view are the building units of the organism, and have a *privileged metaphysical status*" (Soto and Sonnenschein 2005, p. 104, my emphasis).

4.6 On TOFT Systemic Epistemology

Genetics researches supported by development of new technologies in molecular biology have been consolidating the triumph of the *analytical perspective* in biological sciences, which seeks to deconstruct nature to understand its molecular aspects. Saying that natural phenomena are deconstructed to understand the molecular aspects is not the same as saying that 'through our understanding of the molecular aspects we can understand nature in molecular terms'. Let us assume that the debate on reductionism is an epistemological issue that uses an epistemological approach and not an ontological claim. In this context, the term of 'analytic perspective' will be used just in a procedural sense.[19] Stated in these terms, reductionism does not make any assertions on the type of explanation needed for biological systems nor does it emphasize a specific level of investigation of biological reality.

The obvious common use of analytic procedures in the experimental method is distinct from another issue: the (not so obvious) convergence of explanatory models on the lowest possible level of investigation of natural phenomena. The TOFT seems to integrate better an analytic and a synthetic dimension of scientific analysis. The TOFT experimental procedure is reversed: starting from a synthetic approach, it moves to different levels of biological organization using the same analytical approaches that are pragmatically justified. The study at lower levels must be followed by the analysis of higher-level properties, and vice versa, without trying to explain how higher-level properties appeared just in terms of lower level properties and parts. The TOFT has the advantage of proceeding, even if not always with adequate justification from an epistemological point of view, top-down rather than bottom-up. The different schemata, like the analytical and synthetic approach, are not logical alternatives, but they imply one another in the understanding of complex phenomena and especially of biological ones. The "lowest level of analysis" is identified in the tissue. New variables, like cell shape, interactions between epithelium and stroma, etc. can then be taken into account as proximal causes of the neoplastic phenotype.

If the privileged status of the TOFT is linked to its epistemology, such epistemology remains almost always implicit, absorbed into the theorical discussions of reductionism vs. anti-reductionism. In Chap. 5 it will be seen that the mereological perspective adopted by the TOFT in accounting for its own view is a self-imposed obstacle. Abandoning the mereological perspective (Sect. 5.2) would make it possible to go beyond the argument at the right level of explanation and to say in which sense their view is really superior to the other. An attempt will be made to solve these problems with the proposal of a Dynamic and Relational View for cancer research.

---

[19] Let's not ignore that more ample studies of such concepts in philosophical research are possible and also praiseworthy. For the sake of coherence with the present text and the issues discussed, refer to Artigas (1992), Urbani Ulivi (2011a, b), and Bertolaso (2011a).

What the TOFT is supporting is a *theory of fields* in biological sciences rather than a claim in favour of tissues over genes in explaining cancer. Genetic accounts are more frequent in some cancers, for example leukaemia. Tissue structure is so simplified in those cancers that its architectural structure is almost irrelevant, while factors acting directly on cell differentiation play the main role in the origin and onset of this cancer. Cases like these confirm the privileged explanatory status of tissue level and show that the privilege is not linked to the tissues as such, but to the dynamic properties that characterize that and other levels of the biological organization. A systemic view integrates top-down and bottom-up causalities and makes sense of how *sometimes* genes and *more often* cell interactions at the tissue level are causally relevant in the process of carcinogenesis and thus explanatory. The *SMT, when it works (as in the inherited cases of cancer), is a specific case of the TOFT*.

By focusing on different levels of the biological organization in explaining carcinogenesis, the SMT and TOFT do not show incommensurability or incompatibility. They rather exhibit *explanatory independence* while being *epistemologically interdependent*. These terms are taken as introduced by Angela Potochnik (2010). Explanatory independence is "[t]he coexistence of distinct explanations for a single event […]. The explanations are independent in the sense that each individually explains the event in question; indeed, each is the best explanation of the event in the context of certain research interests" (Potochnik 2010, 12 my emphasis). On the contrary:

> By [epistemic interdependence] I mean that the success of these models depends on diverse sources of information about causes not explicitly represented—information gathered with the help of other tools and other fields of science—and that this dependence is mutual (Potochnik 2010, p. 17).

Beyond resounding controversies, the epistemological interdependence between the TOFT and SMT, the idea of TOFT offering the context in which SMT explanations actually work, and the idea of successful SMT explanations as specific cases of TOFT, will all be important pieces of the picture that will be drawn of a Dynamic and Relational View for cancer's dynamics and scientific practice.

Chapter 5
Towards a Relational Ontology for Cancer

5.1 Overview

In this and in the next two Chapters I propose a *Dynamic and Relational View* of cancer to contextualize the explanatory models and aspects of cancer biology. We have seen that Cell-Centred explanations of cancer, by including more and more factors identified as necessary and sufficient conditions, stumble upon a wall of exploding complication. We have considered holistic theories and their anti-reductionist battles, discussing and criticizing the possible reasons for incompatibility or incommensurability that are sometimes thrown into the debate. And we have also illustrated systemic models that, on the one hand, tame complication by transforming it into complexity and organization, and, on the other hand, begin to integrate top-down and bottom-up causality. Now we will shift our attention *from model choice and alternative explanations to the framing of the problem*, proposing the Dynamic and Relational View and a Relational Ontology of biological levels and processes to account for carcinogenesis and cancer dynamics.

Genes and cells – as autonomous parts and causal agents – cannot fully account for cancer (see Soto and Sonnenschein 2005; Baker and Kramer 2007; Bizzarri et al. 2008; Moss 2002), but functional organization of cells in the tissue alone does not account for development and for all the features of cancer either (cf. Chap. 4; Huang and Ingber 2006–2007; Heng et al. 2009). In tumours, cells lose their capability to differentiate properly, but they also end up having a new, specific behaviour that we call 'neoplastic'. Continuity and discontinuity, similarities and differences among parts in the overall functional stability of the systems are difficult to understand without a deeper discussion of (the way scientific practice grasps) *the intrinsic dynamics of biological organizations* and the kind of dependencies that are at work in the organismic dynamism, and that get compromised in the neoplastic process.

From an ontological point of view, I claim that the multilevel phenomenology of cancer reveals the peculiar dynamic organization of metazoans. From an epistemological point of view, I argue that our attention should be directed primarily to the

M. Bertolaso, *Philosophy of Cancer*, History, Philosophy and Theory of the Life Sciences 18, DOI 10.1007/978-94-024-0865-2_5

regulation and organization principles, only secondarily to the single levels that are maintained or disrupted in the hierarchical dynamic organization of living entities. Only in such view we properly interpret why *different levels of biological organization can have causal relevance* for the origin and development of cancer and are all involved in its onset sooner or later. Genes are relevant *as far as* they participate in the organizational structure of biological levels.

5.2 The Organism as a Multi-Unity Dynamism, Not a Parts-Whole Organization

A first fundamental notion that I see emerging from the past decades of cancer research is that a biological entity acts as *dynamic multi-unity,* not as a parts-whole organization. The *maintenance and persistence* of its constitution requires two different modes of causation: (1) a differentiation causal dynamics (multi-) and (2) a state-holder causality (-unity). The differentiation causal principle is mainly related to how the whole is organized through functionally heterogeneous element. The state-holding causality instead is tied to the unitary *dynamic* stability of the whole at different levels of organization.

Unity is an essential feature of any biological system, processual in nature, that is revealed by the way of integration of its organization and growth. Growth, or life history,[1] intrinsically depends on a constitutive and continuous orientation of the parts among themselves and depending on the contextual signals. Adapting biological systems selectively take over environmental signals. The asymmetry so generated is vital in the sense that it guarantees the adequate growth of the organism, as the effects of changes in cell or tissue shape seem to show.

The constant orientation of biological dynamisms entails a tension that takes the form of physical forces and constraints as well (cf. Sects. 4.2 and 4.3). It is such intrinsic dynamic orientation that holds the tension. The identification of the system is not obvious, being it subordinated to questioning what dynamics are relevant in the constitutive changes of a biological system, that is how a constituted system can change and how its parts can interchange without losing their functional and behavioural unity. In fact, the unity we are talking about is *unity of action*. Unity of action exceeds other categories that may seem similar, such as unity of a system or organismal individuality in a traditional sense. Unlike these categories – which are rooted in a mereological framework (see below) – unity of action *admits degrees*.

Whereas parts-whole accounts focus on the characterization of systems from the point of view of their functional structure, the multi-unity account I propose focuses

[1] Life history theory, in ecology and evolution, refers to the timing of key events in an organism's lifetime. Often, the concept of life history is related to evolutionary mechanisms that are responsible for its stability and change, such as natural selection and sexual selection. In this study, we can understand life history as a useful term to stand for growth, development, and more generally the multi-unity history that goes from the birth of a biological entity to its death.

instead on how we can characterize, in a system, structures and functions from the point of view of their reciprocal dynamic and causal regulation. Parts-whole relationships *are to be explained* from the point of view of the kind of regulation holding these dynamics. Indeed, the part-whole relationship might be considered as an instantiation of the multiplicity of dimensions that a multi-unity account entails.

5.2.1 A Dynamic and Relational View of Cancer

Philosophical reflections on systemic accounts have addressed the way in which causality works at distinct levels of organization (cf. O'Malley and Dupré 2005). This aspect of systemic accounts can be fruitfully recovered in a *Dynamic and Relational View* of carcinogenesis that aims at explicitly discussing the reciprocal dependence of the parts and the whole in the structuring of diverse levels of biological reality. Understanding the dynamic generation of biological organizational patterns in nature is the ultimate scientific and philosophical challenge.

The biology of cancer shows that the stability of constitutive elements depends on the organization, and that there is a source of regulation in the biological context. Cells change their behaviour depending on their functional integration in the tissue. Alteration in cell communication alters gene expression, and the loss of integration of cells within a functional tissue leads to genetic instability and apoptosis. *The collapse of levels*, as characterized in cancer, results from the loss of the general functional integration of a biological entity.[2] This means that the structure itself, once constituted, determines the relationships among parts and the stability of the parts themselves (not their final survival). The dynamic structure has no necessary temporal priority over parts (in fact, it appears later than some parts), and yet it has causal relevance, violating a guideline implicitly adopted by SMT and other mechanistic-reductionist approaches: that causal priority is coupled with temporal advance.

More generally, the multi-unity relationship is not properly described by linear causality (including back-and-forth feedback control): instead, we see *a synchronic dependence of constitutive elements' stability* on the maintenance of the organization (cf. discussion of the synchronic and reflexive dependence in Sects. 5.2.5 and 5.4). Indeed, analysing cancer research compels us to consider that *the very definition* of parts and interactions depends on the properties of the multi-unity dynamics. Modular assumptions are, of course, possible, useful and effective in heuristic terms, as part of an idealization process (cf. also Mitchell 2005), but a biological system cannot be considered 'in principle' as an aggregation of its fundamental

---

[2] Note that I have used the expression 'biological entity' and not 'living being'. I believe in fact, that a discussion about what life is or, better, how we should understand it, will be easier after having clarified the relationship between ontological and epistemological issues when looking at behavioural dynamics, which show context dependency and are structured through relational principles in systems that we identify as 'alive' precisely because of such behavioural features. Living beings are paradigmatic of such systems.

parts, as far as the parts themselves are already defined in terms of the higher property to be explained.

The multilevel phenomenology of cancer is easier to understand if we frame the explanatory argument in terms of the intrinsic dynamics holding the biological organization, i.e., in terms of the reciprocal dependency of the parts-whole changes. The heterogeneity of tumour cells can thus be related to the disruption of *relational principles of integration* that hold the normal developmental processes at different scales of the biological organization, and to the intrinsic capability of an organic system and parts to find new functional stable states. As we will see soon (Sect. 5.3), in a Dynamic and Relational View the different kinds of heterogeneity of tumour cells acquire great explanatory relevance in accounting for the neoplastic process (Sect. 1.4.2). Tumours imply loss of regulation of the integrated processes of differentiation at the tissue and cellular level: some conditions are so critically compromised that they destroy the functional hierarchical structure that the organic organization requires and maintains.

Consistently with this view, biological homeostasis is much richer a notion than the maintenance of given functional state, it actually entails a unity of action that links up the discrete and continuum dimensions of biological processes into functional-structural patterns. The only way for a living system to 'survive' is to grow. There is no way to survive maintaining a status quo, which is instead possible in other mechanical and physical systems. The on-going dynamism that continuously shapes the functional integration of levels results in the physiological irreversibility of biological processes. This is the background view of models of carcinogenesis that see cancer as a disease in which sets of molecular pathways for proliferation and tumour suppression or cell differentiation become uncoupled. Some kind of *established synchronization* is lost. In living entities, synchronization is something that changes the time and context dependencies (multi-) and is actively maintained (-unity). It takes the form of *cooperation*, i.e. cells interact in an active and integrated way to maintain the overall functional activity of the body. Carcinogenesis can be considered as a process that occurs in this peculiarly biological time, disrupting growth, that we have also called life history. This is why tissue injuries that are chronologically and spatially related become relevant as far as they create transitions across thresholds that are present in the system's growth. Irreversibility in biological processes is therefore related to this peculiar contingency of biological entities and to how such contingency is structured in their life history, rather than with the impossibility to go back to previous functional states.

At this point, some terms that belong to the semantic area of organizational principles, such as *control*, *regulation*, and *robustness*, need to be differently addressed before going on with presenting the Dynamic and Relational View. The reductionist approach has a tendency to build hierarchies, and to use these hierarchies to identify the level of biological organization that must be studied, as in the systems approach. There is a *relationship of order* (*ordinability*) between levels of biological organization. There is a predominant reference to bottom-up causality in the reductionist view, while top-down and middle-out are often called for in anti-reductionist approaches. *Control* usually implies an element which is external (in the sense of extrinsic) to the system under inquiry, while *regulation* chiefly implies a process whose set point might be internal (in the sense of intrinsic or constitutive) to the system. Both perspectives are possible, although the latter includes the former. These terms emphasize the dynamic component and the hierarchical regulation among different levels of biological organization. The property of *robustness* maintains homeostasis in biological entities but can also be subverted to maintain certain dysfunctions, as in the case of tumour resistance to anti-tumour drugs.[3] Robustness is thus a property which is present and relevant in both physiological and pathological contexts (Kitano 2004b). Cancer, for example, has been described as a robust state that is not optimized for the whole body (Kitano 2004a).[4]

5.2.2 *Operational Integrating Systems*

The multi-unity account of biological dynamisms just presented implies that we describe living entities as *Operational Integrating Systems*, emphasising the functional and genetic stability of the parts and how the organismic dynamism is established and maintained.

Living organisms are systems, where the multiplicity of parts works in a unitary way and where, nevertheless, the stability of parts themselves depends on the progressive amount of structural and functional features in the organism. As said above, the hierarchical arrangement we perceive in biological entities depends upon the specificity of biological organization. The abovementioned unity of action *admits degrees and entangles continuous and discrete dimensions (respectively, processes and, e.g., functional fields)*. Which dimensions come first in explanatory terms is a matter of perspective (cf. Chap. 6). The dynamic network of cells becomes the most relevant issue to be understood in order to explain functional changes at the tissue or organism level. Moreover, if we want to understand the phenomenology of the system-environment interactions in a more comprehensive way, we need to consider how cells and the overall structure of the tissue change *through* such functioning, and how they are preserved, depending on their reciprocal relationship.

The dependence of a functional definition of parts on the whole has been used to support strong emergence (Boogerd et al. 2005), but the argument should be refined in light of the Dynamic Relational View I am proposing. Some views of organisa-

[3] The patho-physio-logical mechanisms involved are regulated by non-linear oscillations of complex signal networks, which operate at the edge of chaos. Mathematical and computational approaches are already developing models that move from a systemic perspective to obtain a unified system of analysis that has useful predictive power (Ge et al. 2003; Phelps et al. 2002; Khalil and Hill 2005).

[4] Robustness is therefore related to organizing principles, it is a relative feature and term, as it is always predicated with respect to something else (cf. Sect. 7.2). Robust behaviours are highly represented in and representative of living organisms, by allowing the maintenance of a certain functional state (physiological or pathological) despite disturbing factors (cf. also Bertolaso and Caianiello 2016).

tion and emergence assume the weak idea of *Aggregative Systems* to account for systemic properties: here, systemic properties are defined by the microstructure of the system's organization. But this strategy becomes problematic when accounting for the kind of biological changes and interlevel dynamics we are looking at. Anti-reductionist positions, instead, often assume the notion of *Non-Aggregative Systems*. "The whole is more than the sum of the parts" – their traditional lemma about parts-whole organizations – directs attention on organizational properties. The functional stability of parts depends on the whole, as we have repeatedly seen, and a weak emergence is in place. The notion of *Operational Integrating System* makes the further step of incorporating the dynamic properties of biological organizations: it maintains that the focus of the explanatory models of cancer should be on how the function and the biological identity of the parts are compromised. When such multi-level dynamic regulation is lost, aberrant heterogeneities of parts (cells) show up and characterize the process of progressive dis-organization that we call 'cancer'. The disruption of such constitutive dynamic organization makes sense of the phenotypic heterogeneity of cancer cells, once the pathological condition for carcinogenesis has settled, and the hierarchical organization has been compromised.

The notion of Operational Integrating System subtracts emergence from the domain of mereology. Mereology is understood as the philosophical theory of parthood relationships. In an Operational Integrating System emergence does not presuppose whole and parts linked by bottom-up and top-down causation; instead, emergence is about causal emergent relationships from which any biological pattern and change has to be understood. Such causal emergent relationships are prior in epistemological terms and justify the use of abstract terms such as morphogenetic fields and functional landscape (see also Aranda-Anzaldo 2002). Outside a mereological framework, the unity of a biological entity is not subject to an arbitrary assumption nor to a yes-no criterion. It comes in degrees. Levels and organismic hierarchical organization are therefore secondary with respect to the constitutive dynamics of the biological regulation, when the question is on development and growth, which also implies that a mereological framework cannot be but derivative. If a relational perspective to account for biological dynamics emerges from the need to integrate in the explanatory accounts the discrete (e.g. cellular or functional fields) and processual (e.g. tissue differentiation) dimensions of the hierarchical organization of living systems, it is no further mereological in nature. This also affects the explanatory framework of biological development, in which our attention goes to the unitary functioning that allows the dynamic regulation of the assembled heterogeneous parts into functional fields or attractors (see Sects. 3.3.1 and 3.3.2).

The effort to deal with the dynamic features of cancer and with the heterogeneity of tumour cells introduces, in fact, a new dimension of analysis in explanatory models of cancer. Such dimension is precisely characterised by the notion of field. The field notion is used with some important differences by Cell-Centred and Organism-Centred Perspectives, revealing the different perspective on the pathological feature of the neoplastic process. In Cell-Centred Perspective the concept of field arises when trying to account for the causal priority of epigenetic events over genetic ones. Clonal heterogeneity of colon or gastric cancer cells is, for example, explained by referring to 'epigenetic' or 'cancerization' fields: these are areas of the tissue in which cells are epigenetically compromised (i.e. altered in their non genetic information). In Organism-Centred literature, instead, the notion of field is constitutive in the explanatory argument. Fields are embodied structural-functional wholes, whose identity – as shown in transplantation experiments – is able to enroll new incoming cells, and is lost when parts get separated (de-part). What is compromised in cancer is not a part which is functionally defined (e.g. neoplastic tumour cells) as in Cell-Centred models, but a *functionality*, described in terms of morphogenetic field. This is why, in scientific practice, the identification of organizational relationships (i.e. networks) is so relevant in explanatory terms, and why so often the attribution of functions to molecular parts has been problematic (see also Sect. 7.2 on this point).

5.2.3 (Strong) Constitutive Emergence

Something more precise must be said on emergence. Mechanistic-reductionist epistemology seems to be compatible with *some forms* of emergence. For example, new cellular activities appearing at various stages of neoplastic progression may be attributed to the fact that the cells (considered as lower level than the tissue organization) have regulatory mechanisms that depend on their context. According to genetic reductionism, in fact, morphogenesis is controlled by the genetic patterning of the body plan, mediated by a cascade of unidirectional and linear gene inductions (Farge 2003). The statement can be suitably amended and supplemented to include mechanical forces, like physical stress, which in turn induce a specific genetic expression, and all this causal chain will be a higher-level property (for example, a collective movement of cells), acting on parts by top-down causation.

Jaegwon Kim (1999, see Mitchell 2012)[5] defined emergence as follows:

> At a certain time *t*, a whole, *W*, has emergent property *M*, where *M* emerges from the following configuration of conditions: *W* has a complete decomposition into parts $a_1, \ldots, a_n$; each a_i has property P_i; and relation *R* holds for the sequence $a_1, \ldots, a_n$. For some a_j, *W*'s having *M* at *t* causes a_j to have P_j at *t* (Kim 1999, 30).

We already know that the multi-unity account is an alternative framework with respect to any part-whole framework. In Kim, we see that emergence still relies on the properties of parts. Conversely, in our analysis there is another aspect which is

[5] Mitchell (2010), while acknowledging that Kim's "views on emergence have set a high standard for clarity in the philosophical community", criticizes Kim's philosophical argument against emergence as conflating compositional physicalism and descriptive fundamentalism.

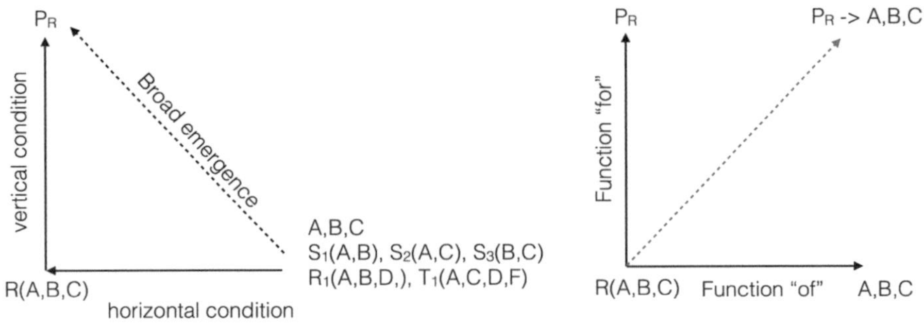

Fig. 5.1 *Left*: Dimensions of emergence according to Boogerd et al. (2005). The vertical condition represents a systemic property P_R of a system R as emergent if it is not mechanistically explainable, even in principle, from the properties of the parts (A,B,C). The horizontal condition shows a systemic property as emergent if the properties of the parts (A,B,C) within the system (R) cannot be deduced from their properties in isolation or in other wholes. *Right*: Emergence in the multi-unity account of biological entities, where the result of the combination of the "belonging relationship" (*vertical*) with the "interacting relationship" (*horizontal*) is the constitutive commitment of parts within the developing biological system R (*violet arrow*).

equally relevant: the biological stability of parts themselves over time. Parts lose their own stability (and hence, properties) once they lose their structural-functional integration in the system. I have argued that the part-whole view assumed by the mechanistic-reductionist stance is not a proper framework to account for biological change and development. A pure mereological account, although it does describe the appearance of level properties in aggregative systems, is not adequate to capture the specificity of this biological organization. Cells constitute an ordered system because of their functional specialization, through their functional and structural belonging to the tissue itself. Parts of a whole do not have prior existence as parts. However, parts belonging to the whole have a causal relevance on their own stability as well.[6] A stem cell in the colon is a cell with stemness properties *because* it occupies a specific place at a given time within the tissue. Its differentiation pathway is strictly dependent on such position in space and time. Tissue renewal properties depend on the stem cell's position and its organized progeny. Outside a mereological framework, the unity of a biological entity is not subject to an arbitrary assumption nor a yes-no criterion. It comes in degrees, and influences our possibility of grasping regulatory patterns in biological systems. Accordingly, we don't need to engage philosophical issues about emergence while relying on a part-whole view. Eventually, in my view, issues with emergence will turn out to be a legacy of the reductionism – anti-reductionism battle. But let us proceed by degrees and clarify what kind of emergence has eventually to be considered.

Boogerd et al. (2005), elaborating on C.D. Broad's notion of strong emergence of the 1910s, distinguish two conditions for emergence, which they think of as "horizontal" and "vertical" (Fig. 5.1, left). A systemic property P_R of R(A, B, C) is emergent if either of these conditions is fulfilled.

[6] This is coherent with other evolutionary and biological considerations about morphogenesis (Gilbert 2006; Soto and Sonnenschein 1999; Biava 2002; Potter 1978).

> The first is the vertical condition: A systemic property is emergent if it is not mechanistically explainable, even in principle, from the properties of the parts, their relationships within the entire system, the relevant laws of nature and composition principles (Boogerd et al. 2005, 135).

The focus of the vertical condition is on whether there is a mechanistic explanation for P_R given the behavior of A, B, and C in R(A, B, C). The horizontal condition concerns, instead, the properties of the parts:

> The second is the horizontal condition: A systemic property is emergent in this sense if the properties of the parts within the system cannot be deduced from their properties in isolation or in other wholes, even in principle. The properties of, say, part A in the context of system R(A, B, C) would be emergent in this sense if they were not deducible from the properties of A, B, and C in isolation or in other systems (*ibidem*).

The two dimensions of parthood[7] are consistent with a multi-unity account of biological entities: (1) the $R \rightarrow A,B,C$ relationship identifies the parts *through their functional belonging* to a biological context; (2) the $R \rightarrow P_R$ identifies the parts *according to their "functionality"*.

The two dimensions of parthood can also be seen as corresponding to two dimensions of function:

(1) a function "of": belonging relationship *F(of)*
(2) a function "for": interacting relationship *F(for)*

Now we can tackle the topic of emergence, linked to the appearance of higher properties in a biological system. To say that "higher properties arise by emergence" means, in our account, that:

> a system *R (A,B,C)* has an emergent property P_R if there is a dependence of the dynamic stability of parts *(A,B,C)* (i.e. their biological identity) on *R* as a whole, P_R as such (functional state), and of the functional role of parts on P_R as an aspect of *R*'s behaviour (intrinsic dynamics of the system).

As graphically shown in Fig. 5.1 (right), the result of the combination of the "belonging relationship" with the "interacting relationship" is the constitutive commitment of parts within developing biological systems, i.e. the $P_R \rightarrow A,B,C$ relationship (violet arrow in Fig. 5.1, right). I take, therefore, *emergent properties* as *physical properties that are identified in terms of inter-level regulatory behaviour*. The dynamic features of the system – i.e., its organization plus its changeability in time – are called emergent properties. They capture one aspect of the aforementioned hierarchical dimension of biological systems: *the explanatory priority of the regulatory aspect over the compositional one*. I call this kind of emergence 'constitutive'. This *synchronic emergence* is a kind of strong emergence which can no longer be considered in terms of a functional dependence of the parts from the whole.[8]

---

[7] These dimensions have been introduced and partially discussed by other authors too, moving from different fields of biological and philosophical analysis (Sober, Sarkar, etc.). I do not believe it necessary to systematically review other accounts considering the present context of analysis. However, further studies will be useful to analyse these convergences.

[8] Boogerd et al. (2005) defend in their paper what they call "synchronic unpredictability". This concept, I think, has its "ontological" assumptions in the *synchronic emergence* described here.

The epistemological status of the dynamic features of the system is related to some kind of causal relevance. We need an operational dimension of causality that accounts for properties *P*. This will be spelled out below. That notion of causality is clearly different in biology than the mode of causality that accounts for property *M* mentioned by Kim (above). We need to associate the two notions without contradiction, with a wider possibility to understand biological systems behaviour.

Parts depend on the whole, but they also maintain – and in cancer, develop – a degree of autonomy. Cells, in fact, keep some basic functionalities (e.g. proliferation) while losing their genomic stability in cancer. But another important aspect of strong constitutive emergence concerns memory. Cells keep track of their own functional career and trajectory within the organism. Stem cells' genomic configuration, for example, largely depends on their biological history in the organism and functioning within the organism. So much so, indeed, that Cancer (Stem) Cells, dislocated within the body, reconstruct a tissue that is similar to the tissue they originally come from. They have a memory, and this memory is physically made concrete in an epigenetic progressive stratification within the cell, characterizing the ongoing regulatory dynamics: so much the part belongs to the whole that its own history reflects the internal history of the organism.

The crucial importance of history for understanding living systems has been well highlighted by a scientist:

> Now it is necessary to introduce a further distinction between dynamics and history. Dynamical systems are described by trajectories in a suitable phase space, but dynamics by itself does not include a distinction between the before, the after and the irreversible changing of the system. As in the case of a gas in a bottle, an observer who plays the reverse motion of each molecule would see nothing unusual, no "before" and no "after". The time arrow, the history and the improbability of events of the reverse-motion film come into play only taking into consideration the boundary conditions. The dynamic approach just associates a clock to the degrees of freedom of a system, but it is insufficient for measuring the global structural changes. To get these ones requires, instead, the observing of the constraints' stratification/modification, which is precisely the history of the system (Licata 2015, p. 47).

5.2.4 Causation "by Holding" Dispels Apparent Circular Causality

Functional states are characterised by the maintenance of two conditions: higher and lower level stability, and functional interactions among parts. Regulation concerns both the functional organization of parts and the stability of parts. It is the life history of the system that requires a focus on the kind of regulation that integrates these two dimensions. Two different causal effects are thus at work in the link between tissue-induced differentiation and cell proliferation. In some explanatory frameworks, these two causal dimensions are grasped by the top-down and bottom-up accounts. In the Operational Integrating Systems that are recognized by the Dynamic and Relational View, the whole doesn't just make the parts: it holds them, revealing a *different mode of causality*. In such mode, effects are not expressed in terms of a progression of events ($A \rightarrow B$), but of maintenance of states (S^*) at different levels of biological organization.

This causal dimension is needed to describe the causal relevance of topologies and biological 'forms' over the maintenance of biological dynamics and over the proper coupling of on-going processes, so that the loss of tissue architecture can be considered by scientists a prerequisite for, and one of the defining characteristics of, the majority of cancers. The host organ ability to normalize cancer cells is influenced by differentiation and developmental factors, as evidenced by the capacity of normal liver environment to normalize cancer cells.[9]

This mode of causality is not about necessary and sufficient factors. The effects of relationships are not 'things' (quantitative phenotypic traits), but *viability conditions*,[10] embodying a concept of *biological necessity* in which the contingency of the natural world is not excluded, nor ignored or problematic. The Relational and Dynamic View implies abandoning the exclusive character of reductionist causality to *broaden the notion of causality* in order to admit non-univocal determinism, whose principles do not follow the logic of what is necessary. The entanglement of causal dynamics is no further linked with the necessary chain of events (cf. the notion of 'program'), but is shaped by what scientists call 'ad hoc commitment': it *implies orientation and functional determination*. The movement under scrutiny is not kinetic (like in traditional mechanistic accounts), nor merely circular, but is the result of synchronizations, coordination and cooperation, fields constitution and rhythms achievement. The effects are not products or molecular changes (e.g., mutations), but *habitats* weaved together as *viability conditions* that embed temporal and spatial information.

Between field and parts, between habitat and cells, there is no causal circularity, but rather *reciprocity with asymmetry*. The pathologic emergence of cancer cells is not simply a matter of a time course (accumulated stochastic events) but reveals a dimension of the relationship that holds – over time – between cells and the structure in which they are functionally embedded. A tissue develops from a long series of interactions during which the cells move, in relation to each other, as part of a process and acquire different identities, depending on these new relationships. From the time when a tissue is formed, the parts we identify can no longer be considered as the parts that had interacted in its formation. The cellular components now present did not pre-date the tissue itself: *they constitutively interact in a mutual way*. In the formation of parts of the body, such as jaws and teeth, epithelial cells together with stromal cells produce the basement membrane, while at the same time this membrane produces the architecture of the tissue (stroma and epithelium). In this context, it is impossible to separate the activity of a gene, or even of a cell, from

[9] Here I am referring to studies by Coleman et al. (1993), Maffini et al. (2005), Hendrix et al. (2007), also showing that the different potential of stromal cells to normalize breast cancer depends on the recipient's age and the time elapsed since her most recent pregnancy.

[10] This distinction avoids confusion with the Kantian notion of 'condition of possibility'. I am using it in a naturalistic sense.

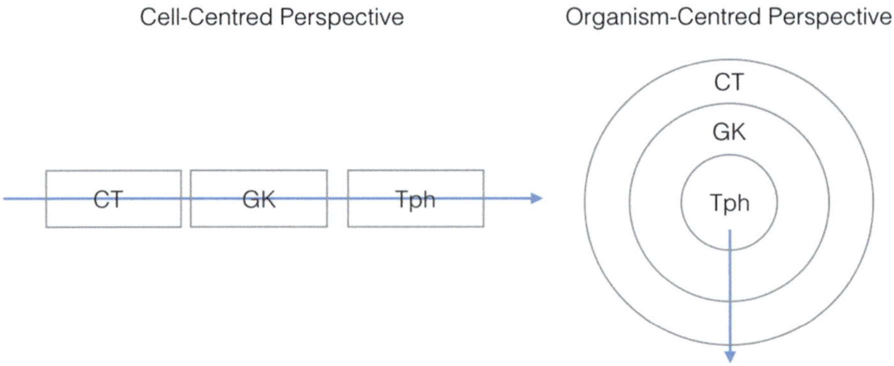

Fig. 5.2 Gatekeeper (GK) and Caretaker (CT) functionalities are differently arranged in Cell-Centred (*left*) and Organism-Centred (*right*) explanatory models to account for the tumour phenotype (TPh) of cells or tissue respectively. The two perspectives converge upon properties that maintain functional states, but their epistemological differences reveal the double level of control captured by a multi-unity phenomenology of biological dynamisms

tissue interactions. A mechanistic temporal dimension (sequence of events) is, in itself, not sufficient to explain how different levels of indetermination may eventually give rise to functional organization.

Such causality "by holding" is particularly apparent in the cateraker / gatekeeper distinction, introduced in the 1990s as an important step in the evolution of (Cell-Centred) explanatory models of cancer. As seen in Sect. 2.5, *Caretaker* (CT) functions participate in the maintenance of the integrity of the genome; they act as the guardians and curators of the system's stability; *Gatekeeper* (GK) functions, instead, have a role in cell-cell interactions and thus cell differentiation, preventing the growth of potential cancer cells.[11]

As it was for "fields" (Sect. 5.2.2), also caretaker and gatekeeper functions constitute a point of convergence among explanatory models of cancer, but with manifest differences in interpretation (Fig. 5.2). The reductionist commitment of the Cell-Centred Perspective works through linear sequences of events, where the causal relevance of parts depends on their temporal priority in the sequence of events, even when the functional characterization describes properties that maintain functional states. Organism-Centred models instead look at cancer as a process

[11] I only have room to mention that there is a third group of genes – beyond caretaker and gatekeepers – whose mutations lead to significant susceptibility to cancer: it is the class of landscaper genes (Kinzler and Vogelstein 1998). These genes encode products that affect and regulate the microenvironment in which cells grow, for example the composition of membranes which, in turn, affects cell-cell communication and the elimination of various kinds of molecules. The disruption of the ECM that may derive from alteration of landscaper genes may determine changes in cell differentiation and proliferation, and even the invasion from incoming cells such as those of the immune system.

where the functional features of fields are compromised, showing a *dual* control upon which the progressive disruption of biological levels of organization depends: a dimension of control is related to the parts' functional organization; the other dimension concerns to their own stability.[12]

The ideas of reciprocity and dynamical interactions to account for change are not new. They were proposed, for example, by the dialectical view, an alternative, anti-reductionist way of thinking and questioning in biology formulated in the 1980s. Geneticist, Richard C. Lewontin and ecologist, Richard Levins are its main authors. The dialectical view emphasizes the provisional nature of systems: change is the underlying principle that characterizes parts and wholes (Lewontin and Levins 2007, p. 120). For Lewontin and Levins, mechanistic-reductionism is wrong in considering parts as homogeneous: elementary, equivalent parts that – eventually – constitute the world. In contrast, the dialectical world view assumes that "(p)arts *make* the whole, and the whole *makes* the parts" (Levins and Lewontin 1985, p. 272). This means that the properties of parts have no prior, alienated (i.e. atomistic, cf. Sects. 2.9, 4.3.2) existence, but are acquired by being parts of a particular whole.

The picture that emerges is characterized in the following terms (Levins and Lewontin 1985, p. 273): change is a characteristic of all systems, because the elements recreate each other by interacting, and are recreated by the wholes they are parts of. Wholes are essentially defined as a relation between heterogeneous parts. There is a dynamic aspect, which is intrinsic and characterizes any complex system, driven by opposition and heterogeneity of the parts within a whole. The dialectical view tries to understand "the causal relation between properties of contextually defined 'parts' and the contextual 'whole' of which they are parts" (Winther 2011; Lewontin and Levins 2007, p. 132).

The systemic approach too raises the issue of "the dependence of the identity of parts, and the interactions among them, on higher-order effects" (Dupré 2010, p. 37). The dialectical view distances from the systemic approach and its static assumptions and explains the relationship among parts and the whole in terms of the generic notions of change and opposition. In systemic accounts the reciprocal dependence among the parts and the whole remains implicit. The points tackled right now could help to improve the understanding of the different directions and senses of causality that hold in Operational Integrative Systems. Like in the dialectical view, parts make wholes and wholes make parts and in this way systems take shape and change through time. But a crucial point of awareness of the Dynamic and Relational View consists in abolishing the *interchangeability between cause and effect*, thus dissolving the apparent causal circularity of all these dialectic, anti-reductionist, and systemic approaches.

[12] This explains the circular scheme of their epistemological position. How this scheme may fall into a reductionist account has been discussed in Bertolaso (2012b) and will not be further presented here, being secondary to our argument.

5.2.5 A Relational Ontology

Anti-reductionist approaches like the TOFT (Chap. 4) explain the causal mechanisms underlying functional organization of complex biological phenomena by appealing to *top down causality*. The tissue as a whole, in the case of the TOFT, influences, and ultimately determines, the properties of its parts (Mayr 1982; Sonnenschein and Soto 1999). The causal circularity so obtained raises metaphysical concerns. As we have seen in Chap. 4, the TOFT tends to put forth a certain level of hierarchical organization – the tissue – as the explanation of a phenomenon such as neoplasy. Yet, TOFT authors point out the problem of circular causality: "How can causation be studied here? Is the tissue causing the formation of a basement membrane? And then, is the basement membrane causing the normal architecture of the epithelium – and thus the tissue? This looks like circular causation" (Soto and Sonnenschein 2006b, p. 373).

A full explanation of this *apparent circular causality* is out reach of experimental science as different hierarchical levels intersect and each experimental approach requires the identification of a level of analysis.[13] From this point on, it is obviously difficult to determine all efficient causes, all cause-effect relationships, as required by an experimental method. "Acknowledging these problems does not seem an exercise in nihilism, but a first step in trying to devise ways of studying organisms while taking into consideration the problems posed by their historicity" (Soto and Sonnenschein 2006b, p. 365).

Organ structure and, consequently, organ function rest on cell types of the organ and on its constituents overall organization, the body. It is the obvious uniqueness of this structure that distinguishes, for example, a breast from a kidney and directs the cells within the first organ to produce milk and the second to filter blood and excrete urine. This is so despite the fact that the cells share the same genome. The processes of tissue functional specificity may thus be extended to organs. The interactions between a cell and its context determine the patterns of gene expression and its differentiated phenotypes in spite of the fact that the blueprint of the genome does not change. "In the end, the unit of functional differentiation is the organism itself" (Nelson and Bissell 2006, p. 288).

If we consider the extension of the causal account we have introduced here, the paradoxical circular causality turns out to be nothing but a projection of the real systemic mode of causation which ultimately corresponds to the principle of functional and structural organization of the living organism and its maintenance (Fig. 5.3).

Historicity becomes the explanatory tool to penetrate the circular causality that is present in the organization of fields, able to treat the apparent "circularity of causes". There is a progressive stratification of biological constraints that track the history of the system. Such a stratification is no further hierarchical. It is historical, correctly captured by the *morphostatic* and *morphogenetic* language (Sect. 1.4.2).

---

[13] In the next chapter I will propose a theory of biological explanation that will employ the notion of "mesosystem" more than "level of analysis". Any studied phenomenon is reduced to the dynamics of a typical mesosystem which is chosen for various epistemological reasons.

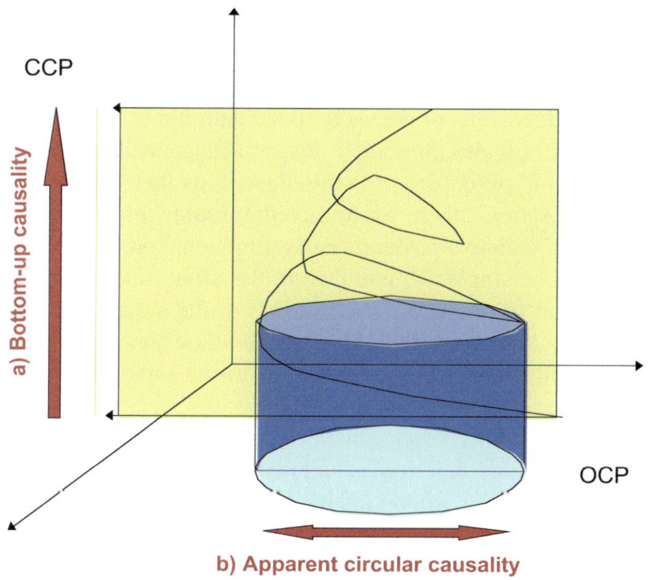

Fig. 5.3 Graphical representation of the different dimensions of causality in the Cell-Centred and Organism-Centred perspectives: (**b**) The Organism-Centred Perspective (OCP) considers the three dimensions of causality: not only top-down and bottom-up, but also synchronic reflexive emergence; in this way, there is no paradox of circular causality, but reciprocity with asymmetry that, in turn, justifies the necessity of adopting concepts such as functional field, and provides a proper observation point on '*top-down*' causality. (**a**) the Cell-Centred Perspective (CCP) flattens the three-dimensionality and spatial-temporal aspects of the biological phenomenon by projecting it onto the vertical plane. The bottom-up causality appears to be the most explicit for integrating the elements and their progressive juxtaposition. The elements, once placed, become essentially interchangeable: whichever plane is used for projection, the same linear sequential structure results

The individuated dynamism provides an order (ordering), but is not given an order (ordered).[14] From the hierarchical perspective, the Operational Integrating account of the biological systems shifts the focus to regulation. *Biological determinations* (see Sect. 6.4.2) depend on their context and admit not only diachronic emergence[15] but a *synchronic and reflexive* one as well (Sect. 5.4).

What follows for the Dynamic and Relational View is that the concomitant discrete and continuous dimensions of biological dynamics and behaviors are grasped

[14] Moreover, if one intends for hierarchy the mere possibility of being ordered, then the neoplastic phenotype should ultimately show an order, and not a disorder (namely the pathological one), accruing the observations of the pathological anatomy of cancer tissue. For this reason, the reductionist paradigm ends up talking about cancer by means of active verbs: cancer is a machine, as the organism in its whole should be, if we follow this same logic.

[15] Diachronic emergence has been addressed by TOFT's authors (Soto et al. 2008, 264) following Bunge (2003) in terms of "the fact that in specific natural or formal systems the initial relations and properties of elements cannot teach us how they would be applied as the system evolves. Thus, the historical way by which a system of natural events operates is not a consequence of its description. It acts and it produces novelty (novel qualities and novel structures) in the real world, which leads to the conclusion that emergence has an ontological meaning" (2003).

by a *Relational Ontology* of biological levels and processes. Such expression emphasizes the synergies of mutual determinations in accounting for biological development and life history. In this framework, the process of co-determination involves different dimensions of causality and ensure the integration of functional macrostates at different scales. So-called "upward causation" is incompressible to a compositional view of "properties of the realizers plus their interactions".[16] Rather than "downward causation", then, we find configurations of relationships that play the role of "semantic" contexts in determining functional properties. As highlighted by cancer dynamics, no single element determines alone, and the failure of process progression is dependent on the progressive loss of the system's integrated regulation. In this sense, a *Relational Ontology* admits dimensions of causality that are asymmetric (e.g., cells do not influence tissues in the same way as the tissue is a relational determining context for their life).

Organismic stability, therefore, can be construed as an ontological property, whose dynamics constantly determines and restores the functional identity of parts in a context-sensitive fashion. Endorsing the perspective of a *Relational Ontology of levels and processes* helps disentangling the difference between the pathological robustness of cancer, overemphasized in the literature (Kitano 2004a), and the veritable dynamic persistence, the one that allows biological systems to orchestrate stability and novelty in face of perturbations and that entails a normative dimension.[17]

A *Relational Ontology* does not override the role of genes and of gene regulation in development. In fact, while cell properties and cell interactions in all organisms depend on the molecules that genes specify, the resulting biological forms and specific cell arrangements are not encoded in a deterministic fashion in the genome (Neumann-Held and Rehmann-Sutter 2006). Functional macrostates are rather engendered by virtue of constraints which, as highlighted by Juarrero, are emergent and dynamic: "Constraints are *relational properties* components acquire in virtue of being embedded in a higher level system" (Juarrero 2002, 133, my emphasis).

5.3 Cancer as Imbalance: Differentiation and Proliferation, Heterogeneity and Homeostasis

In this section I put to work the Dynamic and Relational View and the idea of an Operational Integrative System to interpret some of the most striking features of cancer biology: aberrant differentiation and uncontrolled proliferation. To the question why does it happen that organizational disruption is also connected with the cells' genetic instability, our conceptual framework answers that the cells' instability in cancer is related to a contextual environment that is no longer able to constrain the functionality of parts (cells). The proposed conceptual tools also provide a meaningful understanding of cancer cells heterogeneity.

---

[16] The expression in taken from Kim's view, whose account is clearly inadequate for solving the puzzle of the deployment of "architectural" information (Bertolaso and Caianiello 2016).

[17] See Bertolaso and Caianiello (2016) for an expansion of this argument.

5.3.1 Differentiation and Proliferation

The discovery of human embryonic stem cells (Thomson et al. 1998) and embryonic germ cells (Shamblott et al. 1998), or tissue stem cells present in adult organs like brain and muscles, that can differentiate into different types of cells (Blau et al. 2001), has established the framework through which the whole issue of differentiation and proliferation should be understood. Tissue specificity determines cell differentiation, *thereby* suppressing (or allowing) cell multiplication, or even inducing cell death in extreme cases. As Sonnenschein and Soto noted from a biochemical point of view, stages are probably artificial distinctions because the overall process more likely represents a continuum of biochemical and molecular events, which lead to the transition from a totipotent cell to a terminally differentiated cell.[18] By definition, the process of differentiation requires heritable alteration in the pattern of information of the two cells that are derived from the same parent cell. However, since all the cells of a multicellular organism derive from the fertilized egg, this process must be based on a differential genetic expression in the offspring depending on the process of tissue development. The process must also continue, throughout embryonic development, to give rise to the great – but still limited – variety of cells present in the adult organism.

It is the developmental process that is the critical one in the maintenance of the normal phenotype; when the process of differentiation is compromised, the neoplastic phenotype appears: disorder in cell multiplication derives from an error in differentiation. More properly, the *uncoupling between proliferation and differentiation processes* is the hallmark of the onset of the pathological behavior of cancer cells. One well-described case illustrating this uncoupling is the already mentioned loss of adhesion between cells and the Extra-Cellular Matrix (ECM) (Bissell and Radisky 2001). This dependency is usually indicated by the term "dependence from anchorage". Normal cells that separate from their ECM anchoring system usually promptly undergo apoptosis whereas tumour cells proliferate and spread.

Cancer and differentiation have long been the two great peaks that somatic cells geneticists attempted to climb. Now they seem to have merged into one. In particular, a change in the perspective in oncologic research was determined by the question about tumour latency (see Sect. 1.3.1) and by the discovery that a cancer phenotype could be maintained even after the removal of an oncogene that had initially triggered it, and that the fusion of a tumour cell with a normal one would induce the reversal of the neoplastic phenotype. In this way, tumours are no longer considered the result of changes in the circuits that govern normal cell proliferation, but as "a disease of cell differentiation rather than multiplication" (Harris 2004), so

[18] Embryologists have traditionally denoted the orientation of the cells in a general differentiation pathway with the term cell "determination", reserving the term "differentiation" for the final events through which a cell passes to become a terminally differentiated cell from a pluripotent precursor (Soto and Sonnenschein 1999). I use the two terms interchangeably to refer to "commitment", the common term even in the scientific literature for cells differentiation. I will follow their reasoning here for its biological significance and consistency.

a "block" of critical steps of the normal cell differentiation process is responsible for the onset of cancer itself (Capp 2005).

The empirical data that the discovery of Tumor Suppressor Genes (TSGs, see Sect. 2.3) originated, indicates that the malignant phenotype is a recessive character and that the genetic basis, i.e. the malignity of a tumour, is due to a loss of a normal function which is dependent on the context. As things now stand, it seems possible that the key cellular events determining malignancy are heritable losses of function and in particular, loss of the ability to complete specific patterns of *differentiation.*[19]

Cell differentiation is a vector showing that features of cancer can be removed from their specific biological elementary component (cells) and reformulated in organizational terms. In this way, explanations are not characterized in terms of molecules and cells, but rather in terms of a functional state of the system getting compromised. Such functional state is identified either in terms of a morphogenetic field (tissue in Organism-Centred Perspective), or by attributing organizational properties to the causally identified parts (cells in Cell-Centred models).

Even the "hallmarks of cancer" (see Sects. 2.8 and 4.3.2) can be well understood from a relational perspective that brings differentiation to the forefront, instead of cell proliferation: self-sufficiency in growth signals and insensitivity to antigrowth signals, for example, can be read as *lack of dependence* from growth signals, highlighting the role of cell-cell interactions in defining the functional phenotype of parts. "Hallmarks", in this sense, are the marks that are left behind by the living process as it proceeds in its life history. Indeed, we might say that life is characterized and recognizable by what it leaves behind, not by its parts. This is why talking about physiological and pathological in terms of "hallmarks" is adequate. The "footprint" of life also includes the *places*, the habitats, in which the history of life is in different way represented too.

Summing up, differentiation and proliferation are coupled in the phenomenology of development which is characterized by a *Relational Ontology*. They stick together and allow each other. Instead of focusing on either differentiation or proliferation, the two must be considered jointly. However, there is asymmetry between the two movements: in the process of cell differentiation, interactions show a reciprocity that makes of any 'dynamic in action' (Juarrero 2002) a specific subject of the scientific inquiry. As we have seen in this Section, however, differentiation is more relevant than proliferation in understanding cancer dynamics. Assuming this perspective, the Auto stabilization Model (Section 3.3.1) is of particular epistemologi-

[19] We have assisted to an interesting outcome for Cell-Centred approaches when they take differentiation seriously into account (2....) that we now make more explicit. Even if molecular components are supposed to be responsible for the neoplastic transformation, the concept of cell differentiation that frame molecular component functions adds a *relative* dimension, leaving the self-referentiality – that is typical of the Cell-Centred epistemological stance – behind. That is, looking at cancer with a focus on differentiation forces understanding the very identity of (tumour) cells from a relational perspective – that also includes a time-context dependency – and not surprisingly defined through relative terms as, in fact, TSGs and also CSCs are. Something is differentiated always, in fact, with respect to something else (see more on this point in Chap. 6).

cal interest. The relevant dynamic that follows disruption of differentiation processes is no further mere proliferation but a process of auto-stabilization of the cells within a tissue. Proliferation is paradigmatic just within a mechanistic-reductionist account whereby the metaphorical image of the "renegade cell" takes over the whole explanatory story. Due to the atomistic commitment, the tumor cell (TC) is seen as a sort of cancer (C) in its own right. But differentiation plays the leading role in explanatory accounts, and the relata of the explanatory account necessarily entail functional relationships instead of molecular identification (see Chap. 6).

5.3.2 Homeostasis and Heterogeneity

Another aspect of cancer is the loss of the balance between homeostasis and maintenance of functional heterogeneity. From a morphological point of view, in cancer there is *a general loss of order and functional heterogeneity*. Data suggest how cancer genesis (Malins et al. 1998), and in particular the behaviour (Coffey 1998) of solid tumours (Calin et al. 2003), are a chaotic deterministic process characterized by reduced complexity, as indicated by the loss of determining "golden means" and the disappearance of "self similarity" (Sedivy 1999). In the transition from the proliferative to the differentiation phase, cancer progression brings the primary population of cancer to a degenerate stationary state by means of altered gene expression dynamics, loss of connectivity and collectivity (Waliszewski et al. 2001). These peculiar characteristics bring to an increase in instability and to the loss of *ordered heterogeneity* at a genetic, structural, temporal and functional level (Rubin 2007; Posadas et al. 1996).

As we have seen several times in the preceding chapters, numerous studies show how cancer cells can be normalized once they are in a normal environment (Mintz and Illmensee 1975) and how the cells maintain a certain capability of differentiating, notwithstanding their genetic alterations (Kenny and Bissell 2003; Lotem and Sachs 2002). It has been demonstrated that genetic instability can be induced when cells are cultivated without control in a microhabitat (Maitra et al. 2005), and the same process is probably induced in pre-malignant cells, following the destruction of effects at the tissue level. Consistently, the Dynamic Reciprocating Model exposed in Sect. 3.3.3 studies the importance of features such as polarity to maintain tissue stability and the phenotypic specificity of cells. Moreover, recent studies have evidenced how the asymmetry of normal cellular division is also lost during the process of stabilization of the neoplasm phenotype (Cicalese et al. 2009), reinforcing the idea that a characteristic feature of the organismal dynamism is polarization (cf. Sect. 4.3) and orientation in a given environment which are, therefore, semantically relevant.

Such pervasive polarized phenomenology through different levels of biological organization is relevant in the articulation of the argument in favor of synchronic reflexive emergence (Sect. 5.4): multi-unity dynamisms are oriented and polarized in a given environment. The subsequent deregulation of the maintenance of genetic pathways, generated by alterations of the microhabitat, suffices to generate genetic defects, such as the ones observed in cancer cells. Therefore, the mutations that inactivate the specific genes, involved in cell differentiation, can be considered as a consequence of one or more mechanisms, due to the *loss of function*.

The initial condition of carcinogenesis consists, in sum, in an imbalance between functional heterogeneity and homeostasis at different levels that hold the system's functioning states.[20] If cancer biology highlights the pivotal role of homeostasis in fostering the integration of functional levels in the organismal system, the *unorganized heterogeneity* exhibited by cancer cells illustrates the failure of the veritable systemic regulation at the organismal level. System homeostasis should be considered as a concomitant force in the maintenance of differentiation, so that progressive tumor cells heterogeneity might be clearly tied to different levels of (genetic or epigenetic or metabolic) instability of the system. In this sense, cancer can be considered as the price we pay for evolution, the latter being understood as the coupling of the differentiation process that maintains and foster functional heterogeneity of the system and its homestatic dynamics. Analogously to what happens with sexual reproduction mechanisms, that limit genomic alterations in order to prevent macroevolution, the multiplicity of homeostasis levels constitutes the limit the system implements to prevent macro-evolution at the systemic level (Heng et al. 2008, 2009).

Homeostasis at multiple levels is increasingly considered as an important factor to control cancer by the scientific literature over the last two decades. This is consequential and unsurprising. In accordance to this, robustness of a network, reversibility of the properties of epigenetic regulation, tissue architecture and the immune system play a more important role than individual genetic alterations (Soto and Sonnenschein 2004; Jaffe 2005; Feinberg et al. 2006; Heng et al. 2006a; Harris 2005; Huang et al. 2002; Martien and Abbadie 2007).

5.3.3 *Organismic Plasticity*

The problem of *plasticity* is also overturned by this analysis of cancer from a perspective of uncoupled processes. Systemic approaches (Chap. 3) showed clearly how the malignant phenotype can become normal without changing the genotype, in contrast with the Cell-Centred Perspective, in which plasticity was seen as a property genetically acquired by the cells (SCs and CSCs). "Thus phenotype can be dominant over genotype; *(s)ignaling pathways are context dependent;* (m)aintenance of homeostasis requires maintenance of form" (Nelson and Bissell 2005,

[20] Heng et al. (Sect. 3.3.4) observe that "Interestingly, when each layer of homeostasis is broken down by cancer cells, the genome contexts are different from the constrained cell populations" (Heng et al. 2008, 2009). Even in the case of resistance to pharmaceuticals, the new cells present altered kariotypes. In this case, new systems are formed from the kariotypic heterogeneity that breaks the limits of therapeutic pharmacological treatment.

p. 300, my emphasis). Plasticity is a passive default property of cells and of a living system: on the basis of extracellular signals, which are internally mediated by the various molecular and genetic pathways, plasticity of the differentiated states allow for the evolution of the phenotype throughout the life of the cell, the tissue, the organ and the organism, in order to secure its survival capability. This phenomenon is attributed to the differentiated phenotype, presenting at the same time characteristics of: (a) *robustness* or stability with respect to minimal perturbation (reason for which a breast can never become a kidney in vivo) (b) *plasticity or reactivity* to external perturbations.[21] What needs to be explained, then, is not plasticity, but the cell and tissue phenotypic stability: how this is constituted and maintained in different biological systems. Differentiation is, in fact, a process fully integrated to the morphogenic one, so the question becomes what level of functional stability or specificity is compromised in a disease like cancer.

Knowledge of development, homeostasis, cancer and ageing converge within a biologically coherent and significant framework. The concept of tissue plasticity and cell differentiation, so important in our view of cancer, can be seen in the light of developmental biology. Any organ structure contains an informative component different from the genomic blueprint, and performs an integration of signals. The different tissues have notable differences both in their composition and in their general architecture, and they clearly present different functional specificity. It follows that it is the tissue microhabitat that directs the development of an organ and tissue specificity. Once again, cancer research can be interpreted as pointing to the abandonment of a "moleculecentric" view, as Nelson and Bissel argue in the following passage:

> Organ architecture is thus both a consequence and a cause for development, differentiation, and homeostasis. But how does the architecture of an organ (or tissue, or cell) make itself heard? We understand something about the alphabet (ECM, receptors, cytoskeleton, nuclear matrix, chromatin) and even less about the rules of grammar that turn random words into commands (activation of tissue-specific response elements). We believe that decoding this language requires abandoning the currently fashionable "moleculecentric" style of inquiry and adopting a more interdisciplinary approach that takes into account dynamic changes, spatial segregation of events, and tissue architecture (Nelson and Bissel 2006).

As Nelson and Bissel observe, "When one considers all of the signalling pathways involved in differentiation, the complexity is staggering. There is clearly more than one way of integrating the same combination of signals into a phenotype (…) this is precisely why development is so miraculously robust" (Nelson and Bissell 2005).

---

[21] Throughout pregnancy, for example, the breast faces significant transformations in view of the function it will have to carry out after birth, and aggressive carcinoma cells can be reorganized in order to form normal tissue, changing their microhabitat or simply changing the microhabitat signals (Weaver et al. 1997).

5.4 Relational Categories

Cancer biology can really contribute to the understanding of (physio)-pathological phenomena, because it provides elements for a deeper understanding of living entities. Cancer is not as a set of molecular parts nor a mere process, but a disruption of *the ongoing relational interactions that constitute* an organism.

The starting point is considering the organism not as a hierarchically organized system, made by parts and wholes, but primarily as a multi-unity dynamism. While one causal principle promotes differentiation, another principle – a state-holder one – holds structural unity and organization. The emphasis is on identity of organisms and systems. Hierarchical structures and boundaries are not elementary but derivative from this dynamism. Indeed, they admit degrees and breakdowns (as in cancer). Freed by the primacy of mereology, we become open to explore the dynamic generation, maintenance, and disruption of biological configurations. The part-whole view – or "mereological" view – is a self-defeating starting point if we have to study living systems and processes, i.e. the *unity of action* that characterizes biological entities. Finally, an operational perspective implies considering not only how parts are organized in a system, but also how this organization is maintained, i.e. to focus on the *modes of regulation* at work in the organization and change of biological systems. A variety of descriptions of the multi-unity dynamism is, in fact, possible. This justifies the multiplicity of non-reductionist models we have seen in Chap. 4 as well.

In a Dynamic and Relational View, I have proposed the notion of Operational Integrating System, a concept to understand biological functional states and the way in which they are established and maintained through change. Other notions – e.g., Aggregative System and Non-Aggregative System – are imbued with mereology, and do not allow to focus on the modes of semantic regulation, and on how the stability of parts depends on the whole. The Operational Integrating System – a system ordered by integrative principles – which is object of inquiry is there in spite of the changes and movements of its parts. The 'peculiar way' in which parts are assembled is *relational*[22]: in physical terms, long-range interactions and topological features are more relevant than mechanistic properties of the organized parts. The principle of functional integration is the System itself. In Operational Integrating Systems not only do we have kinetic movements, but the *consolidation of patterns* of significant structural and functional sense. This pattern gives identity to such system, by defining and integrating it into functional fields, the organism's configuration. In the neoplastic process, where all that is compromised, phenotypic heterogeneity highlights the stochasticity on which all physiological processes rely.

[22] Such 'ordinability' is linked with relational causality. I have called elsewhere Bertolaso (Bertolaso 2012a) this feature 'cardinability' to emphasize the concomitant presence of discrete and continuous dimensions in these dynamics. However, a clarification of this point would require the discussion of other empirical and epistemological issues that have not been explicitly presented in this volume.

The very stability of parts depends, synchronically, on the whole stability as the whole existence depends on the parts although, as we have seen, a crucial asymmetry holds between parts and whole, since any part needs to appeal to the broader organization for many aspects of its definition. Parts are defined and explanatorily relevant for their semantic dimension: the baseline is that there are no such things as isolated parts, and that, when they interact, they exchange something because there is reciprocal "acknowledgment", and change along the process. There is a dynamism rather than equilibrium or mere flow, and this is captured by the notion of *synchronic reflexive emergence*.

I have suggested that the relation between Caretaker and Gatekeeper systemic functionalities can be usefully captured by the multi-unity account of biological dynamics proposed above. Conceptually, they resemble the morphogenetic and morphostatic mode of causation. In the organismic processes we have been considering, genes are therefore explanatorily relevant as far as they enter the organizational structure of biological levels. No problem, then, to admit that genes can have a causal role, but this is not related to the simple assumption that they entail the whole program to structure the organism, or to disrupt it. Their role is causally relevant as far as they can be characterized and functionally defined, on the basis of the systemic properties that eventually account for the tissue level of biological organization. Interactions maintaining the structure and connection within the organization, though physically weaker, are crucial in maintaining parts' stability and functionality as well. These forces are "organizational" and are realized through different kinds of physical and chemical interactions. Again, the mode of organization has a causal effect on the functional structure (cells' organization), and is causally relevant on the molecular level as well.

The parts-whole organization, as long as it lasts, needs to be maintained by the different dimensions of causality at work. The inadequateness of "what follows by necessity" requires a revision of the intrinsic structure of causal accounts. Necessary conditions only hold in an ideal system, while systemic theoretical entities (e.g. functional fields, attractors, etc.) are closer to the real world captured by science, than the genes that played such a relevant role for a long time in Cell-Centred cancer research. A wider notion of causality might be at hand. The disruption of the equilibrium between the different modes of causality for constituting and maintaining a living entity and the uncertainty of the ways in which the neoplastic phenomenon may arise, characterize the relationship between causality and uncertainty in the Dynamic and Relational View. It seems that science suffered from a long philosophical tradition that reduced causal issues to a mechanical relationship between input and output, and is thus challenging the causal framework that holds kinetic movements. The causality "by doing" is complemented, in the Dynamic and Relational View, by a causality "by holding". The dialectical view in philosophy of biology had already postulated the effectual reciprocity between parts and wholes: parts make the whole, and the whole makes parts. In the Dynamic and Relational View, "make" has different meanings in the two parts of the sentence. We not only have reciprocity, as in the dialectical view, but *reciprocity with asymmetry*.

The account counters the great emphasis on cell *proliferation* as the primary feature of cancer, providing a picture in which *differentiation* is primarily impaired along with the rhythms and long-range interactions through which organization is maintained. Proliferation, in this framework, appears as subordinate to differentiation (including apoptosis and other coordinated behaviors of the cell). Anti-reductionist theories stated that the architecture of the tissue is an emerging property of the cellular society and not just a simple function of the collective properties of the cells that constitute it: its causal and explanatory relevance cannot be reduced to properties of the parts. The Dynamic and Relational View inherits the organism-centred consideration of cancer in terms of impairment, or uncoupled interactions, among cells and between them and the tissues, and the existence of morphogenetic fields, that enable and preserve specific interactions like conveying positional and historical information. The process of cell differentiation is regulated by several internal and external signals, but always in a contextual manner, relative to the reference system. So much so, indeed, that even Cell-Centred models end up into paradoxes while trying to keep the focus on any specific context-insensitive feature that would determine the proliferative behavior. It is the differentiation program that is the critical one in maintaining a phenotype. Speaking of cell differentiation means including scaling issues in considering the mutual dependency between cells and their contexts, which is *intrinsically relational*. The causal link between differentiation and proliferation capacities is relational, and when the differentiation program is lost, neoplasy appears.

The properties of a biological system, from the point of view of its behaviour, are determined by its specific identity (relationally, not essentially defined) and as a consequence, a certain degree of uncertainty may follow. The multi-unity account of biological entities makes sense of the different kinds of heterogeneity observed in cancer (see Sect. 1.4). It explains the different kinds of dependencies at work, that are characterized, at the phenomenological level, by the physiology of reciprocity with asymmetry and polarization, and, at the level of pathology, by the progressive increase of heterogeneity within tumors.

The important intuition behind the new perspective has to do with the ability to focus on the relationships themselves, described in terms of morphogenetic fields. The Dynamic and Relational View, in some way, comprehends reductionist, systemic, and holistic approaches, relating them in a new, pluralistic and productive interaction. Cancer theories that look at the organization of cancerous tissues as the main causal factor of the disease (see Chap. 4) do not suggest to look for objects that cannot be taken into account by the molecular perspective. Instead, these theories suggest considering objects and processes from a new point of view, which is much more comprehensive. The advance is conceptual. With respect to systemic and anti-reductionist theories presented in the previous chapters, then, the Dynamic and Relational View makes a new and deeper reflection on the *nature of observables*, arguing for a *Relational Ontology* of biological phenomena and entities. The foundation of this ontology lies, indeed, in constructing and constituting *relationships*, rather than on parts and wholes or "levels". All concepts (e.g., emergence) have to be intended in this new way.

Chapter 6
On Biological Explanations

6.1 Overview

In this chapter I sketch out a theory of explanation tightly linked with the notion of Operational Integrating Systems, and, more generally, with all the dimensions of the Dynamic and Relational View of cancer explored in Chap. 5. We will deal with a fundamental epistemological duality constituted by, on one hand, the identification of the explanatory level according to our research interest and, on the other hand, the characterization of the system whose dynamisms we want to study. We will introduce the idea of "mesosystem" (between micro and macro) and its derivatives, arguing that reduction operates by "mesoscopic reasoning", seeking the right explanatory level for the dynamic that needs to be understood. We will use the mesoscopic concept to reflect on why all cancer research seems to be converging on the tissue level and to derive some criteria for choosing a particular explanatory level in scientific practice. Reduction is always possible, and often fruitful, but it is constrained by validity conditions that are directly determined by the relational nature of the studied systems, whose elements are functionally defined by the higher level properties and ontologically dependent on them too. This is also why anti-reductionist views can play the role of defining the systems and the contexts, while reductionist views can't. We will also deepen the crucial notions of "stability" and "specificity", and "determination" and "indetermination". The two pairs of terms are fundamental to appreciate how our theory of explanation is deeply entangled with the operational nature of biological dynamisms.

M. Bertolaso, *Philosophy of Cancer*, History, Philosophy and Theory of the Life Sciences 18, DOI 10.1007/978-94-024-0865-2_6

6.2 Two Distinct Epistemic Dimensions: Identifing the System and Structuring the Explanatory Account

Human reason is able to sail in the deep waters of biological contingency and complexity, bringing into the same account two epistemological dimensions.

The first dimension has to do with the identification and choice of the explanatory level or level of analysis, by characterizing the system that holds the dynamic behaviour of interest. The second concerns the definition of the *relata* that structure the explanatory account.

If we consider this double epistemological dimension, we see that any biological explanation entails a non-reductive dimension which is evident not only in the definition of the system, but also in the identification of explanatory parts.

As shown in other studies too, the non-reductive identification of parts is interestingly exemplified by network studies in Systems Biology. Palumbo et al. (2007), for example, demonstrate that a couple of mutations involving two enzymes in yeast, which per se are not essential, cause death of the organism *if* the double knock-out provokes a "lack of alternative path" condition in the whole metabolic network. Here, two concurrent non lethal events acquire an essential property, lethality, from the existence of a global metabolism architecture, not by some deep internal 'nature' of the two enzymes. In other words, their lethality is a collective emergent property of the network system (Bertolaso et al. 2013; Giuliani 2010).

Topological issues are crucial: in network studies, all the properties relative to each node are derived only from its pattern of relations (edges) and thus from its peculiar location in the complete graph (Palumbo et al. 2005, 2007). But this *'essentiality-by-location' principle* holds for any reduction that is performed in studying complex problems in biology. Collective effects of cellular behaviours – that are so relevant in cancer – cannot simply derive from the knowledge of the constituting elements.

The epistemological relevance of the relational features of the biological systems asks for such articulation of the explanatory process. The real challenge in biology, and I would say in science in general, is related to the human capability to frame the question so that relevant entities, or levels of generalization, emerge in the explanatory enterprise.

The network paradigm in Systems Biology constitutes a unique synthesis between reductionist (all is in the molecules) and holistic (all is in the whole) approaches. It shows that various parts-whole explanatory frameworks and the top-down-bottom-up causalities can be seen from an integrative, instead of oppositive or dialectical, perspective.

Still, a constitutive asymmetry remains between the assumptions of the reductionist and anti-reductionist (or organicist) perspectives. The asymmetry directly derives from the two epistemological dimensions mentioned above: identification of the system and characterization of the *relata* of the explanatory account.

Anti-reductionst claims are particularly adequate to the definition of systems. Anti-reductionism, beyond any specific model proposed (e.g., a holistic top-down model), may be seen as drawing an attention to the definition of the system and of the parts. Reductionistic commitments, instead, are not able to play at this table. In fact, deep contradictions come up when "reduction" is considered as a principle that is able to define both the system and the structure of scientific explanation at once (and in the same way). The definition of biological explanatory entities resists reductionism, leaving the impression that something other than mechanistic-reductive requirements is needed. This kind of reductionism is generally "impassible" (Bertolaso 2013a) for the explanatory challenge that inter-level regulatory features pose. Reductionism is good for the second epistemological dimension, the one concerning explanation. In that dimension, we can also tolerate the doubt about the existence of "new key entities" that might explain something more about a phenomenon.

The different capacity of reductionism and anti-reductionism to access the two distinct epistemological dimensions (explanatory account and system characterization) is an asymmetry that needs to be recognized and kept firm in order to obtain a unified picture of the interpretation of the neoplastic process inclusive of the debates and the tensions generated by apparently opposed views of cancer.

6.3 Mesoscopic Style of Reasoning

When facing complex multi-level biological phenomena, the accuracy of an explanatory account does not only depend on the level of details obtained through different technologies: the interplay and reciprocal dependence between the scientific question and the phenomenon to be analyzed is also crucial. Different levels of explanation are possible for the same phenomenon, but they are not usually answering the same question. Not all contexts and functional activities of parts are equally relevant to a specific scientific question. Information theories suggest that the selection of the level that controls the system is crucial for a study that wants to have significant implications in terms of predictability, whilst the most accessible level from a point of view of information results is not always useful to control the system. As Waddington says, a deeper comprehension of how complexity and information theory can be applied to bio-systems will clearly determine research strategies (Waddington 1977).

In cancer research, as years go by, we see different approaches converge on considering the tissue as a privileged explanatory level. What cancer research is showing with this convergence is, I think, that the intrinsic features of the explananda are able to drive the convergence of different explanatory models towards the same level, without an a priori definition of a privileged level of causal explanation.

Tissue properties are capacities or *viability conditions*: they create an environment in which the underlying dynamics can take place. This is why it has been said that the architecture of normal tissue consists of 3D organizational units that comprise the morphogenetic fields of the organism and are the carriers of positional and historical information. But the choice of the tissue level depends less on the level at which the pathology is observed, the histological one, than on the fact that the causal biological relationships that hold that specific functional level explain what is actually going on in the process of carcinogenesis.

We see that what determines the level to be 'preferable' follows the resolution of scientists to get to an adequate explanatory level of the observed phenomenon. *This style of reasoning is named 'mesoscopic'* because it concentrates on the relation structure that is considered as the channel relating the microscopic elements and macroscopic parameters. It implies the necessity to take into consideration, even when we concentrate on a single element (e.g. the lethal character of a specific mutation), the general functional frame in which the element is inserted.

The mesoscopic level is where "organizational principles act on the elementary biological units that will become altered, or constrained, by both their mutual interaction and the interaction with the surrounding environment. In this way and in this place is where general organization behaviour emerges and where we expect to meet the elusive concept of complexity" (Bizzarri et al. 2011, p. 176).

The identified system constitutes a level of order or *mesosystem*[1] and exhibits regularities (*determinations*) where not only the properties of the parts from inferior levels emerge, but also the peculiar relation established among the parts of the system themselves.

6.3.1 Reduction as Identification of the Mesosystem

Reduction coincides with the identification of the mesosystem, whose explanatory value underlines the heuristic value of the system's perspective. If reductions in science are not meant to explain the world, but to give us pieces of knowledge about it, while defining the way we get those pieces of knowledge through empirical research, most concerns about the partiality of reductions disappear, and the dichotomies between reductionist and non-reductionist features of scientific explanation can be overcome. Models don't even try to explain all the aspects of a given phenomenon. Models are much more ambitious in trying to grasp some features of a phenomenon to control it in a proper way, often depending on pragmatic interests. The experimental practice does follow this path when first identifying stable functional behaviours before discussing the specificity of the interaction of the parts that constitute the systems that show that behaviour.[2]

Explanatory accounts are open to integration (between different explanatory models depending on the instrumental tools adopted), they don't need to fear new

[1]The roots of this concept is found in the field of Systems Biology and of the methodological considerations presented by Noble (2006).

[2]Relevant considerations with respect to this issue have been also presented in Buzzoni (2015).

attempts to explain a phenomenon, because their explanatory driving force is not based on the kind of proof adopted, but on the kind of question to be answered. Their real challenge is in terms of generalization in a different field of scientific inquiry, not the apparent threat that can come from the same one. Instead, this kind of worry threatens the work of those adopting mechanistic-reductionist perspectives apparently more open to changes in scientific research.

Consider Woodward's claim:

> depending on the details of the case, description or causal explanation can be either inappropriately broad or general, including irrelevant details, or overly narrow, failing to include relevant details. Which level is most appropriate will be in large part an empirical matter (Woodward 2010).

However, in a relational account, what an "empirical matter" is should be understood in a wider sense. It is not just a practical problem (what is possible here and now) but a rational problem that starts with a scientific question (why this behaviour and not another one).[3] Pragmatic reasons in scientific practice are embedded with the effort to identify mesoscopic levels where objective and subjective dimensions of science meet. And the structural and functional dimensions of the system cannot be separated. They are captured differently by the definition of the system and by the functional behaviour of the parts.[4]

The scientific question and the related choice of the working level influences both the results obtained and the form of the explanation. A well-posed scientific question determines the focus of the research, together with the theoretical elaborations of the experimental data. From this point of view, the analytical and synthetic approaches converge into a Dynamic and Relational View that is capable of integrating experimental data with the theoretical working hypothesis formulated at the level of biological complexity with the highest degree of explanatory coherence. A form of 'non-trivial determinism' is, therefore, at stake (see also Bertolaso et al. 2013).

Both molecules and collective dynamics are equally involved but this does not necessarily imply that they are equally explanatorily relevant, in the sense that they play a different explanatory role in the process of understanding. We have historically been called to study neoplastic processes through an epistemological and methodological approach that breaks cancer down into simple elements, identified as the main causes that, once isolated, can be explanatorily sufficient. Molecular biology, in this way, provided a suitable experimental platform for this kind of sci-

---

[3] This opens a philosophical reflection about the correspondence between the world and the way we know it through science. A compositional and pluralistic view of the scientific enterprise is better suited to explain why and how science works, but the philosophical implication of what is a common practice in science does not seem to be at hand yet.

[4] Moreover, describing functional dynamics that are nomologically dependent on the context through models, where those dynamics are reconstructed in mechanistic terms, allows us to make machines able to perform those functional properties through parts and devices that act in a mechanistic way. As noted by Agazzi, this possibility does not imply equivalence between living and non-living systems. The latter remain nevertheless 'artificial' because they are not able to undergo all the reciprocal relations in which natural organs are usually involved (Agazzi 1978).

entific work. As mentioned earlier, however, empirical evidence forced the revival and resurgence of a more comprehensive understanding of the phenomena, which implies a synthetic approach besides the analytic one that characterizes the experimental design and represents and important feature of scientific practice as a rational enterprise.

This methodology can be generalized to include other situations by defining 'windows' in both space and time: under this heading the tissue organization in the process of carcinogenesis is a very convenient mesoscopic scale because it maximizes the determinism of the occurrence of tissue organization disruption. Therefore, to some extent, the normative dimension of biological explanations is settled by the experimental design: we *reconstruct dynamics* by means and errors, through models that always give us partial answers on the dynamics, as different regulatory levels interact and operate in a synergistic manner, requiring a more comprehensive explanation than a mechanistic one. The search for the level that maximises non-trivial determinism must be intended as a step to find 'where to start the investigation'. Its primacy is in terms of the descriptive enterprise and not in terms of the particular piece of world. On the contrary, the regularity and functionality of a level may be dependent in different ways on those at a lower or higher level. This is clearly the case in many biological morphogenetic phenomena of cellular differentiation or pathogenesis, such as cancer. These processes show that in nature there are relational factors that play a crucial and directional role.

6.3.2 Conditions for Reduction

The need to bring emergence into the picture is not a general claim about the 'in principle' impossibility of reductionism in science but questions, instead, the specific requirements for any actual reduction when regulatory features of biological behaviour have to be explained. I see no interest in advocating 'in principle' impossibility of reduction in science. What just seems unavoidable is an analysis of its conditions of validity or (its) requirements. We can reasonably say that 'reductionism' has equipped us with important results because of an epistemological privileged status of genes, albeit considered as functional elements more than molecular 'parts', with all the consequences of such account.

The conditions for reductionist explanations are thus the conditions for any scientific explanation looking for mechanisms that, of course, always exist at any level of the biological organization. However, acknowledging that mechanisms are always in place is not a claim about the 'reducibility' of biological organizational features. This is what I gather from 'system level understanding' of biological phenomena, and I suggest this perspective as an aid to the philosophical understanding of scientific practice, a help to understand better what kind of results are important, and for what.

An account of scientific explanation fitting well with our view is Kenneth Schaffner's *Preferred Causal Model System (PCMS)* account. A classical thesis in philosophy of science is that the reduction of one theory or one branch of science to another one, considered more fundamental, is possible. Schaffner (2006) started off from criticizing such account of reduction, arguing that it is not applicable to scientific practice in biology. Schaffner introduced an analysis of biological theory as a *collection of overlapping causal and inter-level models*. In 1993 he distinguished the Nagel type of generalized reduction-replacement model (GRR) from *a causal/mechanical approach* (CM) that works through *partial reductions*. The partial reduction of the CM approach (paradoxically, in Schaffner's view) is typically multi-level in both the reduced and the reducing sciences. These reductions are partial because scientific explanations always deal with, or include, higher-level features while accounting for them in terms of molecular features and mechanisms, so that these reductions mix higher entities and predicates with relatively lower-level entities and predicates.

Two elements formally characterize causal mechanical explanation models. These are Field Elements (FE), referring to plausible explanatory candidates (generalizations, mechanisms, kinds of experiments, etc.), and Preferred (Causal) Model Systems (PCMS) that, in Schaffner's account, have to be understood as "causal system[s] representing a temporal process" (Schaffner 2006, p. 387, 1993) (Fig. 6.1).

In Schaffner's account, PCMSs implicitly or explicitly involve laws and generalizations relevant to the scientific problem at hand. They are causal and qualitative, describing parts of mechanisms in a process.[5]

There are three conditions for a successful partial reduction in terms of CM explanations: (1) the *explanans* is identified with parts of the organism or a process of interest – this usually implies the assumption that the parts (or processes) are at least partially decomposable microstructures of the system under analysis; (2) the *explanandum* (i.e. event to be explained) is a grosser (macro) typically aggregate property or end state; (3) assumptions that permit relations of the macro and micro descriptions are specified. The latter are called connectability assumptions (CAs), or bridge laws and reduction functions.[6] CAs can be causal sequences or identities, which are related with some regulatory aspect of the system of interest. Reductions are thus causal mechanical explanations that require these three conditions and make use of FE and PCMS.

However, one of the major features of biological processes is precisely the temporal *disengagement of causes and effects or a midstream interruption of the cause*

[5] In this way, Schaffner makes explicit his original concern about the nature of theory in biology: while incorporating mechanistic explanations in his CM approach, he still takes into account the possibility that the real nature of biological theories might have been originally misconceived, avoiding the discussion of their nomological character or resolving it in mechanistic terms.

[6] As Schaffner does in his paper (2006), I will also consider this point in its epistemological dimension. Logic and ontological discussion are also required and already developed, at least partially, elsewhere (Mitchell 2010; Bertolaso 2013a, b) but are beyond the aim of the present volume.

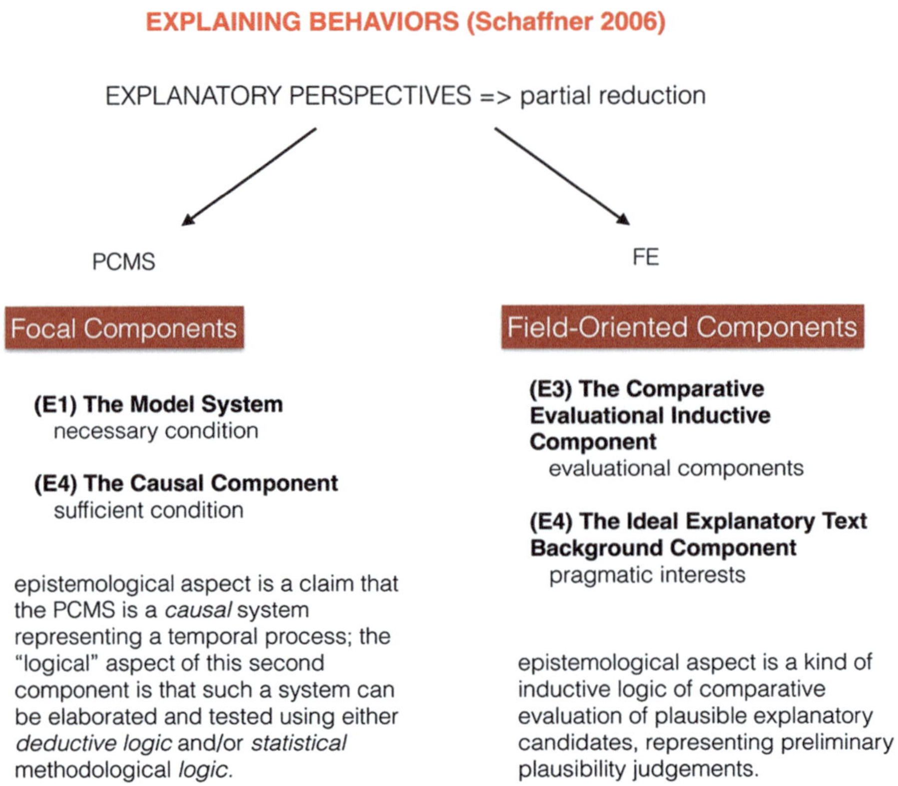

Fig. 6.1 Model of reductive explanation developed by Schaffner (1993, 2006)

while the effect still holds. This is one of the intrinsic features beyond the idea of inter-level regulation. Problems arise when ignoring this point and expanding the explanation beyond the limits that the *explananda* had posed. Strong mechanistic accounts involve a regular set of changes by definition. This also implies a *continuum* of events, given that the entities and their relationships have to be maintained. The requirement of (temporal) directionality of causal accounts, as understood in scientific explanations, is also intrinsic to the latter condition. One of the differences between mechanistic accounts and the CM proposed by Schaffner is that the latter does not require such continuum among different levels of the biological organization. At the same time, it has the formal tools to grasp such requirement (disengagement of causes and effects) in biological explanations. Its epistemology is more articulated and distinguishing between epistemological dimensions that follow a deductive logic (PCMS) and an inductive one (FE), leaves a space open for the dimension of the process of scientific explanations that require a comparative process. CA identification belongs to this aspect.

Schaffner's account of partial reduction should be understood in epistemological terms, as a respose to the "failure of any possible explanation of a whole in terms of its parts and their relations (and expressed only in the parts' language)" (Schaffner 2006, 382). On the other hand, the account might allow adopting, as a first approximation for an epistemological analysis of how reductions work in science, a pragmatic approach that acknowledges our current impossibility to derive higher-level behaviours from the knowledge of relationships among systemic elements. This approach leaves room to the general concern about reductionism in biology about the possibility of explaining higher-level properties in terms of interactions of parts that take place at a different, lower, level of the biological organization. An "innocuous emergence" is commonly accepted, i.e. the thesis that parts do not tell you what the whole will do without a specification of the interrelations among the parts themselves.

In partial reductions, a different epistemological role should be acknowledged to the parts and to the system, by distinguishing, in the process of reduction, different steps to identify the level of generalization and potentially admitting the dependence of the parts identification on the higher-level features, through the definition of a PCMS.

The process of identification of a PCMS is meant to allow apparently unrelated features of the parts of a system to be reorganized to the extent that some of them can be assumed to be 'proximate causes' within a given context. These relational features can subsequently be considered in mechanistic terms. *Once the system and its functional parts are identified*, a question about the dynamics and how parts interact is always possible. However, the enterprise of defining the PCMS is not set and resolved within the mechanistic framework. Moreover, the partiality of any model and of its causal account might eventually be related to the kind of generalizations required to account for biological processes and their pathological features. Given this point of view, some of the conditions that Schaffner lists among the non-reductive elements of CM are to be moved into a different, more adequate account of reduction. In this way the PCMS is a model of reduction that works within a wider number of scientific enterprises, i.e. not only among those interested in the behavior of parts within a given system (typically molecular biology), but also those interested in the system's overall dynamics typically captured by disciplines like Systems Biology. Reduction implies a reference to *'middle-level' entities*. It is a process that identifies causally qualitative, higher level models because of the normative dimension entailed in the kind of generalities characterizing the definition of the PCMS.

Schaffner recently coined the useful term "creeping reductions", as opposed to "sweeping reductions":

> I call 'sweeping reductionism' where we have a sort of 'Theory of Everything' and there is nothing but those basic elements – for example, a very powerful biological theory that explains all of psychology and psychiatry. The second kind is 'creeping reductionism' where bit by bit we get fragmentary explanations using inter-level mechanisms (Schaffner 2013a; see also Schaffner 2002).[7]

[7] In the same article (2012), Schaffner is moving from coupling the term of "creeping" with "reductionism" to coupling it with "reductions". The reason is the difference he makes with sweeping reductionism and the focus on the plurality of potential "creeping reductions" that can be performed and are exemplified in his paper. I adopt this useful distinction.

An explanation is reductive as it appeals to entities that are parts, but *it is also non reductive because*:

> (1) it does not explain all the cases of the scientific problem under consideration, (2) it is plausible that key entities have not been yet discovered to account for the higher level behaviour of interest, (3) it refers to *middle-level* entities (mainly collective systems), (4) it is a *causally qualitative model* (not quantitative and only roughly comparative), (5) the set up of the models takes place at *a (aggregative) level higher* than molecular (rephrased from Schaffner 2006, 397–398).

Schaffner asserts that a creeping reduction is not committed to a "nothing but" account of the (biological) world, while relating explanations to inter-level mechanisms. He leaves open the discussion in which sense and to what extent mechanistic explanations really fit with the inter-level regulation of biological systems.[8] For more details on these issues about mechanism and reductionism see the Appendix.

Acknowledging the role that the identification of a PCMS plays in creeping reductions helps us to understand the relationship between the pragmatic aspect of modeling and the validity of the elements that contribute to the explanation of the process as well. Thus, an anti-reductionist claim is mainly related to this epistemological dynamic: there is no privileged level at which multi-level phenomena can be causally explained (something supporters of a mechanistic view of science would agree with), but the identification of the relata of any mechanistic explanation is dependent on the levels' properties and other explanations can be available in non-mechanistic terms, focusing on different aspects of the emergent properties. In this case, different perspectives are possible, some more useful than others, depending on the practical aim and purpose.

Finally, we should take into account that the notion of the PCMS would probably need to be extended to handle robustness issues. One way of doing it would be to consider the notion of "pathways", widely used in (systems) biology. As Schaffner defines, "a pathway is a coordinated causal sequence that may contain entities at different levels of aggregation […] with a defined endpoint, which may be a behavior or facet/component of behavior" (Schaffner 2016). Studying robustness would require using a number of alternative PCMSs with flexible shifts among their activities. These shifts might depend on interruptions in one path that shunt the entire system into another path. All this route of philosophical reflection still has to be pursued (Schaffner, personal comm., 2016).

There are thus degrees and thresholds that can enrich the picture, but it is necessary to accept that the process of generalization of the explanatory terms can change depending on the issue at stake (e.g., genetic accounts explain less than CSC

[8] On this point some considerations about the kind of "information" that is required in Schaffner's definition of strong emergence can be developed: "all the information about the parts and the connections will never allow an explanation of the whole" (Schaffner 2006, p. 383) along with a discussion of the caveat about interrelations among parts included in the last part of the definition of "innocuous emergence" (ibidem, pp. 382–3).

accounts the rate of cancer spreading and the temporal relevance of epigenetic factors, etc.). What mesoscopic approaches and related *partial reductions* allow is precisely to distinguish the epistemological roles that parts and contexts play in biological explanation.

6.4 Stability Wins over Specificity

In philosophy, stability and *specificity* have been largely discussed along with other notions, like level of explanation, though without getting to a unified account of their role in scientific practice (cf. Woodward 2003, 2006; Lewis 2000; Yablo 1992; Mitchell 2000, 2008; Noble 2006). Close to Woodward's definition (2010), which meets the most general use made by the experimental design, mainly I have referred to *specificity* as the feature connected to the extent to which a causal relation approximates the ideal of one-cause-one effect. Instead, *stability* has to do with whether a causal relationship continues to hold under changes in background conditions.

The Dynamic and Relational View entails that, in the process of scientific explanation, *specificity is secondary to stability*. In fact, only where functional states and properties are relatively stable is specificity possible. For example, only in states where the karyotype is relatively stable (e.g., in development), can genetic mutations and epigenetic adjustments play a dominant role. The stochastic evolution of cancer, instead, makes it impossible to establish direct causal relationships with *specific* genetic or epigenetic features. In a word, there are no specific functions, allowing explanatory specificity, without functional stability.[9]

In the first chapters of the book I showed how, over the last decades, difficulties related to the *multiplicity of causal factors* challenged the *specificity of the causal role attributed to molecular parts*. Also, the *contextual dependency* of the tumour cell phenotype forced a consideration of the relevance of some established dynamics that take over the control of the tumour cells' behaviour. Cancer research has moved towards models that are either integrated by many different molecular parts, or capture dynamic features of the neoplastic process, through systemic approaches. Yet, an unambiguous definition of cancer has always been difficult because of a peculiar causal complexity of the disease at the organismal level. As a consequence cancer biology became increasingly linked to its dynamic and processual components, putting some functional aspects at the centre of the construction of interpretive models.

Initially, the firm goal of cancer research was the search for key mechanisms and elements (e.g., genes) that, *being specific*, could become the target of treatment. In

[9] On this point see also footnote n. 10 of Chap. 5 and the relevance of the structure of molecular biology experiments.

this predominance of specificity, in vitro cultures remained the privileged experimental system, to some extent favored by the long-lasting impossibility to study single cells and by the difficulties to deal with the whole organism. In vitro cultures were long considered sufficiently homogeneous to assume that all the units contained therein were (causally) equivalent. Any result was considered in terms of average values and functional properties of the cells defined in this way.

The Dynamic and Relational View takes on the accumulated evidence that cell lines, established in vitro, do not offer a suitable experimental model, as they reduce the complexity of the (alleged) proliferative phenomena observed in vivo. The equivalence between cell culture results and those obtained in growing animals, which are ultimately a reiteration of the phenomenon to be understood, can often be regarded as incorrect.

The reconstruction of the functional context of the tissue microenvironment provides a key condition for tests where specificity can be studied. A stabilizing role has been therefore progressively attributed to contextual factors, which include long range interactions and topological factors. Robustness of networks, reversibility of the effects linked to epigenetic regulation, tissue architecture and genomic analysis have been gaining importance, while computer simulations have described the relational conditions in which disorder in the morphostatic gradients generates the precursors of epithelial cancers in the stroma, in absence of genetic mutations (Baker et al. 2009). In this way, the modeled organization of normal tissue and the progression of morphogenetic change linked to diffusion phenomena, show how the destruction of morphogenetic gradients is sufficient to provide the aberrant cell phenotype. The cell is freed from *gradient-based control*, irrespective of the presence, or absence, of genetic mutations in cancer cells, during the initial neoplastic process. Basically, we can say that the architecture of normal tissue is a 3-D organizing system that, like morphogenetic fields, carries positional and historical information. The cells have a memory system so they know where they come from: their history, and where they are; their position, and the integration of this information restricts and circumscribes the future differentiation and movement options open to them. Both association patterns and cell types change as tissues and organs are formed. Consistently, the immune system too plays more important roles than identified genetic alterations (Soto and Sonnenschein 2004; Jaffe 2005; Feinberg et al. 2006; Heng et al. 2006a; Harris 2005; Huang et al. 2002; Martien and Abbadie 2007).

The stability/specificity entanglement in our explanatory perspective offers a way to capture the pathologic feature of tumour cell behaviours. Difficulties of reconstructing discrete stages and the impossibility to attribute the origin of cancer to a unique intracellular molecular component or specific exogenous factor, can be revised from the point of view of emergent functional states. That is, during the neoplastic process the molecular components are mainly unvaried, but *their functional activity* (*functionalities*) changes, due to internal and external factors that eventually involve multiple DNA-damaging events as well. Such change is considered dis-functional as far as it does not respond to the normal regulative factors properly (e.g., aberrant differentiation) and brings about a change of the subsystems as well (e.g., genetic instability). From this point of view cancer can be considered a disease of the on-going systemic organization of an organism, of its natural dynamism. Parts lose their integrated functional properties and become more rigid falling into apparently functional states that mainly require a lower level of energy to be maintained.

What follows from what we have been presenting in this Section is that *to be specific* is a challenge related to the identification of an adequate mesoscopic level, i.e. the identification of a level of biological complexity at which a functional stability is lost. This correct identification becomes a key prerequisite for the development of the model. Defining the dynamics at stake in the maintenance of a stable functional state has a methodological and epistemological priority over the identification of the (dis)function of specific parts. In this epistemological sense, then, *stability wins over specificity*. The neoplastic phenotype is no longer defined on the basis of molecular characteristics, but in terms of functional states that are compromised and disrupted.

6.4.1 The Functional Test

The "stability wins over specificity" perspective is also interpretive of the numerous studies showing that cancer cells can return to normality when placed in a normal microenvironment and maintain their ability to undergo apparently correct differentiation, despite genetic defects (cf. Chap. 1; Mintz and Illmensee 1975; Hochedlinger et al. 2004; Kenny and Bissell 2003; Lotem and Sachs 2002). The changes in the genome would then be causally specific only in the context of global destabilization of gene expression.[10]

As seen for the genomic approach (Sect. 3.3.4), during the stochastic evolution of cancer it is impossible to establish a direct causal relationship between environmental and specific genetic/epigenetic factors, like for many processes described in the literature on morphogenesis and development. As already mentioned above, only in states where the relatively stable karyotype does not change can genetic

[10] On this point of particular interest is the work done by Capp (2005). Special attention has been also devoted to the demonstration that genetic instability itself (therefore, the accumulation of mutations) follows the onset of an abnormal microenvironment, as studies seem to demonstrate the genetic instability of stem cells, when grown without control of the microenvironment (Maitra et al. 2005). The same could happen in pre-malignant cells, after the loss of the stabilizing effects from the organization of surrounding tissue. The subsequent deregulation of the DNA maintenance pathways, generated by alteration of the microenvironment, would be sufficient to generate the defects observed in cancer cells, so mutations that inactivate specific genes involved in cell differentiation may be, more generally, a consequence of the other non-mutational mechanisms, prompting the remark (already found in Chap. 1 regarding cell differentiation) that, "It may be more correct to say that cancers beget mutations than it is to say that mutations beget cancers" (Prehn 1994). Here, however, I focus on the context-dependence of the *effect* of mutations (i.e., specificity) once they had occurred.

mutations and epigenetic adjustments play a dominant role, similar to what occurs during adaptive micro-evolution. From a systemic point of view, karyotypic changes represent "points of no return" in the evolution of the system. Moreover, *When the system is not homogeneous* stochasticity acquires particular relevance. A population approach is more adequate than ergodic ones. The collective overall dynamic takes over the activity of the single cells (the role of stochasticity is inverted with respect to what was assumed in the Cell-Centred Perspective). This is consistent with the search for *correlations* in genomic and systemic approaches, under certain initial conditions, rather than linear *cause-effect connections* between molecular components of the system in order to establish a causal relationship between different functional-disfunctional events.

All these and other bunch of data (cf. Sect. 1.4) are, therefore, particularly consistent with the experimental relevance of the *functional test*. This test is required in order to confirm the tumorigenic properties of tumour cells. At the level of in vitro cell cultures, there are well known potential confounding factors that justify that the final experimental test must be an in vivo test. Among other factors there are, for example, the differential growth rate among subpopulations in culture, and evidence that some behavioural traits of cells in culture, consistent with some characteristics of many tumour cells, are also encountered in cells in normal culture, so they do not constitute proof as such. This is the case for microscopic changes, so that most in vitro changes towards the neoplastic phenotype are reflected primarily by cytological changes, like membrane and signalling differences (Ruddon 1995). Some macroscopic traits can be added to the list, such as proliferation changes in culture, sometimes referred to as "immortality": transformed malignant cells become "immortal" in that they can be passaged in culture indefinitely. This list of traits should also include ability to grow both in a less rich medium, and to higher cell density, than those required by non-transformed cells. Other traits frequently seen are the ability to grow in soft agar, along with the ability to ignore signals to either stop dividing or to undergo apoptosis.

The ultimate demonstration that a modification is carcinogenic is, however, the ability to show malignancy in vivo: the ability, upon transplantation, to produce tumours. The definition of a cancer cell, identified through biochemical and molecular criteria, always requires a functional test in vivo. Many of the features described above are incidental to the phenomenon as they either do not affect nor are affected by it, or may be consequential or secondary manifestations of it. Other qualitative characteristics are required to identify a tumour as such. This also explains why, eventually, functional characteristics are those that identify tumour cells in the Cell-Centred-Perspective as well (cf. the hallmarks of cancer, Sect. 2.8). Patterns – regularities that allow us to define the right causal level at which a biological phenomenon can be explained – take the form of stable functional states, of functionalities.[11]

---

[11] On this point cfr. also the analysis offered in Bertolaso 2009a, b of cell-centred and systemic account of the hallmarks of cancer.

Conflicts between the explanatory relevance of the system and its environment can hardly be resolved. The point is that molecular or structural elements, per se, are not sufficient to explain the phenomenon of cancer (in its phenotypic characteristics). To put it in another way, when facing inter-level regulatory (or context dependent) processes, a different causal relationship between necessary and sufficient conditions is required. Two causal dimensions (the system and its context) have to be brought into the explanatory picture. Necessity – as we usually understand it, i.e. as consequentiality of causes and effects – is not adequate to make sense of what happens in these biological processes. What is required, instead, is a comparison between functionalities (or capacities) where relative terms are represented by the cellular and contextual elements.

6.4.2 Biological Determinations

The theory of explanation inspired by the Dynamic and Relational View of cancer fits well with a notion of "general determinism", so that a *determination* does not have to be defined by quantitative variations or by external movements or by being unique and well-defined:

> All that is needed in order to maintain determinism in a general sense is to hold the hypothesis that events happen in one or more definite (determinate) ways, that such ways of becoming are not arbitrary but lawful, and that the processes whereby every object acquires its characteristics develop out of preexisting conditions (Bunge 1979, p. 13).

The more appropriate notion of 'determinism' to investigate dynamic phenomena is, therefore, not related with necessary and sufficient conditions. The main philosophical explanatory concern about biological processes is more related with the nature of *biological determinations* in general, than with ideas of top-down causality or emergence. A *biological determination* could be understood as a set of regularities that hold, under certain initial conditions, a dynamism identified through a relational account. The Dynamic and Relational View of biological behaviours admits the concomitance of causes and effects that have the character of *viability conditions* (cfr. Chap. 5).

No essential properties and necessary and sufficient conditions are required to make sense of the peculiar dynamics of biological determinations. There is no need to advocate for key entities, the mechanistic interactions of which should be causally specific for the phenomenon. *Relational categories* with their own analytic dimensions are required.

The very possibility of identifying and studying these determinations implies recognizing a causal dimension that systemically builds up the dynamics of organisms and promotes the establishment of *new properties*. We have seen that the appropriate notion of cause is not merely bottom-up or top-down, but synchronically reflexive (causation by holding, Sect. 5.2.4). The discussed *circular causality* (cf. Sect. 5.2.3) is apparent, being a projection of real and dynamic, generative, causal dimensions that can be recognized in living entities. To be treated through the scientific method, its deterministic effects must be analysed at a given point in its evolution and historical organization. The sectioning of the phenomenon on one level, instead of another, allows the identification of a bio-logically significant intermediate system (mesosystem), in which the circular component can be studied in terms of feedback and forward mechanisms too.

The complexity typical of living entities could be defined as 'order in indetermination'. The nature of this concept of indetermination has much to do with the relational principles discussed for biological determination. Rather than arising problems related to predictability (typical of the debate between determinism and indeterminism), it forces a reflection about the ontology of biological organizations and the most adequate epistemology to adopt, in order to understand them. Consistently, biological indetermination is understood differently by reductionist, anti-reductionist and systemic perspectives (Chaps. 2, 3 and 4). In a Cell-Centred Perspective, the biological indetermination that characterizes the different levels of the biological organization can be either ignored or reduced to noise (eventually eliminated by using a higher number of samples). On the other hand, to think in systemic terms, means to always couple the process with the constituent elements of the system, so that contingency becomes explanatory in itself. The determinism[12] underlying the reductionist perspective is that of the program, genetic, epigenetic as may be, but always a program, an element that identifies the system and essentially determines its intrinsic evolution of parts. On the other hand, the determinism characterizing the anti-reductionist perspective is functional (cf. Bertolaso 2013b). In the Dynamic and Relational View, instead, the concept of "indetermination" stands to the concept of "determination" as the "concave" to the "convex" and not as "white" is to "black", i.e. like a term is to its opposite.

The concomitance of top-down and bottom-up causality and of the intrinsic indetermination of the organic systems with their deterministic processes, is not an argument against the determinism of the organic structure and of its functional stability. The combination of a stochastic dimension of parts behaviour and of the local determinations of the system's properties, guarantees the plasticity of the system as a whole and the possibility to explore new functional states at local levels.[13] From this point of view, stochasticity must be considered as a source of that cell heterogeneity that characterizes tissue functionalities and, when such functionality is disrupted, of

[12] Despite the wide-ranging debate on the question of determinism, in this study we refer to this concept only for those elements that can provide better understanding of the philosophical doctrines dominating also scientific research programs in the last century. No deterministic doctrine is a consequence only of the observation of phenomena, it is also, and above all, the result of a number of conceptual assumptions (Ferrater Mora 1994) that are not indifferent in scientific practice.

[13] The topic would have to be looked into in greater depth and from a broader perspective considering how, through analogous mechanisms, living systems can evolve and structure their "knowledge" of their functional state (phenotypic identity) and their environment. This would also be necessary in order to understand better how the informational dimension, contained in the genes, and that which pertains to the biological context, relate to one another. This point goes beyond the objectives that we have proposed here. It is a topic that is open to reflection concerning the peculiarity of the growth and development of living organisms.

tumour cells' heterogenity. It is not mere 'noise' but 'variability', a sort of intrinsic *indetermination* of any biological system, that the system itself requires to stabilize its functional state or to explore another one.[14]

The *unsaturated dimension* that characterizes any biological entity allows for different explanatory accounts. Biological uncertainty is, therefore, related with the openness of the natural world so that both the definition of the observable and the causal categories adopted are dependent on such interplay between the observer and the object under inquiry.

The temporal and spatial components of the neoplastic process are also embedded in the notion of functional fields when specified in morphogenetic terms. In such morphogenetic fields, "functional" and "structural" correspond to two different dimensions of the spatial-temporal realization of *dynamism in action*.

6.5 A Relational Account of Cancer

The Relational Ontology proposed in Chap. 5 is inseparable from a suitable theory of biological explanation. According to such theory, as we have seen, decomposition and composition are not trivial procedures in scientific practice. Knowledge of *biological determinations* is reached by means of *mesoscopic style of reasoning*: there are causal relationships that are caught by the mesosystem and by the process of its identification. Biology describes *differences*, and in biology understanding is always achieved *by relating*, not by isolating supposed causal factors. In fact, the identification and study of any part (gene, cell, cell type, tissue etc.) requires *relative terms* and entails *relational issues*.

We have seen that the Relational Ontology emphasizes that a biological system rests on the properties and interactions of its parts while preserving a degree of plasticity by not completely constraining the stochastic behaviour of the parts and creating new levels of stochastic behaviours as well. There is a sort of indetermination of the biological material that justifies the plasticity, but it is not this that generates the order, which rests within the hierarchical system and that, once compromised, manifests itself through the structural and functional disorganization of its parts. Indetermination is the condition for this generation of order. *Its relevance belongs to the mode through which order is generated.* This is the different dimension of causality we should consider, when addressing questions related with biological behaviours.

The limits of the reductionist approach are not so much tied to the determinism that the theory stubbornly tries to hold on to, as they are to the interpretation of (the role of) causality. The determinism of the mechanistic-reductionist perspective

[14] Such difference, that is crucial from an experimental point of view, can be further clarified analyzing different kinds of noise in biological sciences and how they are treated (cf. Bertolaso et al. 2013). It is interesting how this point meets the concern about the limitedness of the natural selection argument made by Vineis et al. (2010, cf. Section 2.8).

states that material events (processes, entities, etc.) are completely and satisfactorily explained by their prior physical states of matter or events. But a deterministic perspective is also compatible with a Systemic view, so some "properties and interactions of the parts are determined and thus explained by the properties of the containing system, that is, [...] at least some of the relevant explanatory properties of the parts are not independent of the system in which their bearer is to be found" (McLauglin 2001).

The proposed framework reconceptualizes reduction as the identification of the *mesosystem*, i.e. of what is *explanatorily relevant*. Such identification follows pragmatic reasons, but also convenience with respect to the phenomenon to be explained (the *explanandum*). Not by chance, all experimental and theoretical approaches to cancer tend to converge upon the tissue level, i.e. the level where determinism is maximized. In any case, reduction is always *partial*. In Kenneth Schaffner's words, what we have are creeping reductions that operate by the identification of Preferred Causal Model Systems. This is how the Dynamic and Relational View overcomes the dichotomies between reductionism and anti-reductionism.

Reductionism is not "passible" in biological sciences: it necessitates some conditions. First of all, there must be stability which, in turn, allows for specificity. Specificity means a privileged causal role, and specificity is what reductionism looks for. But only in states where emergences of the larger context are relatively stable is specificity possible. Tumour heterogeneity underlines the loss of specificity of the parts with respect to the whole and, in a certain way, also the plasticity of living matter that is organized through morphogenetic processes. This impairs, for reductionism, the "passibility" of many of the imaginable Causal Model Systems. In particular, it prevents the possibility of strict genetic determinism. This is the reason behind the claim – advanced in Chap. 4 (Sect. 4.5) – that in the (rare) cases where the Somatic Mutation Theory works, it does so as a specific case of the Tissue Organization Field Theory. In other words, it is the tissue, with its configurations, that sometimes creates the conditions for genetic determinism (and genetic reductionism) to work.

A very important distinction in the emerging theory of explanation is the distinction between two epistemic dimensions: the *definition of the system* and the *abstraction of some properties of the parts to explain* features of the system's overall behavior. Reductionism runs along the second epistemic dimension, but it stops very soon when the definition of the system is at stake. On the other hand, the identification of parts and the abstraction of their properties require a definition of the system. This is why I argue that *any biological explanation entails a non-reductive dimension*. The properties or behaviour of a (fully defined) system appear, in a certain way, uncertain because of the intrinsic properties and interactions of the parts (see also McLauglin 2001). It is not only a matter of considering the parts in a teleological perspective, but to assume that the real system acts on its own parts establishing relational causal interactions. This way of structuring found in biological systems goes together with the establishment of a *habitat* that is carrier of biological information. This point could be developed into a revision of the proper notion of biological information. In my opinion, what Waddington said should be taken much more seriously:

> If a better idea of the real nature of complex systems is arrived at, if they are not thought of in terms of static functions of the quantity of information contained in them, but rather would that a more dynamic question be asked over how much instruction have been necessary for its realization, or that the instructions tend to impose themselves on their environment (cf. C.H. Waddington 1977, p. 145).

Chapter 7
Complementary Issues of a Relational View of Biological Determination

7.1 Overview

A satisfactory explanation of a complex biological phenomenon like cancer implies different questions that are simultaneously present: some are related with the dynamic organization of the system and others with the definition of the system itself (cf. Sect. 6.2). Such double dimension is always present when considering regulatory patterns in biological systems. Both kinds of questions seem to be answerable in terms of "function", an incredibly sensible term both in scientific practice and in philosophy, object of incredibly many conceptual inquiries. Statements of function are implicit explanations (Gayon 2006), but different notions of "function" might answer different kind of questions. Functional explanations seem to suffer by an intrinsic circularity that equally requires philosophical reflection and whose epistemological nature needs to be clarified. We will capture the notion of function which seems more fruitful to account for cancer research. Conversely, cancer research offers an interesting case study in order to shed light on some philosophical issues related with functional account and theories of function.

All such epistemological "equipment" will be put to work in Sect. 7.2 to analyse the apparent conceptual paradoxes of Cancer Stem Cells research, showing that these paradoxes are dispelled by looking at CSCs from within the Dynamic Relational View of Cancer.

Even for the reductionists there is a metaphysical price to be paid for functional attribution. The price is the acknowledgement of the real existence of the system (entity) which acts as the principle of integration of its functional parts. The definition of stem cells, for example, leads to a biological definition of the cellular component that is inherently context-dependent, defined temporally and spatially. The system is conceived, not as a juxtaposition of elements, but as a set of elements in which the *relative position* acquires a peculiar relevance, as it defines parts functionalities in space and time.

M. Bertolaso, *Philosophy of Cancer*, History, Philosophy and Theory of the Life Sciences 18, DOI 10.1007/978-94-024-0865-2_7

When this level of conceptualization of a biological system is not reached, even biological functions disappear. Functional attributions take the form of activities, that in the SMT are considered as typical of a molecular element (genes or cells), i.e. as belonging to a biological individuality as such. Active verbs that refer to a biological individuality are justified: a tumour cell does, proliferates, metastasizes, etc. Functional explanation, instead, stresses another aspect, i.e. the fact that such biological individuality has some properties as an instantiation of a wider class of biological identities having that property in their own right, i.e. primitively or essentially, but sharing it with other subjects as well.

The teleological problems that arise from functional discourse in biology do not derive from illegitimate appeal to final causes, but rather from an apparent holism that involves relational causes.

The evolutionary argument on cancer, based on natural selection at the cell level, appeals to functions such as proliferation, production of specific antibodies, etc. The famous statement by Dobhzansky, "Nothing in biology makes sense except in the light of evolution" (Dobhzansky 1973) is shared by both the reductionist and systemic perspective in the literature, but with some important differences: the concept of evolution is not the same. For the former, evolution is a mere progression driven by chance. Fot the latter, it is the significant integration of elements and parts driven by the interplay of dynamic behaviours, or coupling of intra-levels dynamics, able to create new functional fields or contexts. The real complexity of the neoplastic process refers to the characteristics of dynamic components that require, for an adequate understanding, analysis of the regulative features of the emerging nested structure.

7.2 On Biological Functions

As we have seen in the first chapters, functional characterizations of genes and cells are present all the way up in Cell-Centred models: genes and cells become explanatory insofar as we are able to answer what genes do, what the tumor cell does, etc. The Dynamic and Relational View I have proposed postulates two different dimensions of the explanatory structure that are mutually implied. One is related with the *dispositional functioning* of parts within a context, the other with the *functional definition of biological systems* that operates through its constituent parts. We are dealing with two interrelated aspects of biological functions that can be seen as descriptive tools for the constitution and decomposition of dynamic systems. Therefore, "the basic aim of function talk is not to explain the occurrence of a part in a system, but to identify the system as a whole and analyse it into functional components" (Toepfer 2012, p. 6). To what extent such view of biological function corresponds to traditional views – the *dispositional* and *etiological* theory of functions – and to recent organizational theory of functions is the object of analysis of the following sections.

7.2.1 Theories of Function

The Etiological Theory[1] of function argues that the assignment of function to a given feature has no meaning except in relation to the past history that has led to its existence. The Etiological Theory leads to searching for the origin of biological functions, and it exploits the heuristic potential of identifying a system based on its functional properties. The Dispositional Theory[2] of function makes the reverse choice: it is interested in how a system exists in a specific time and assigns function to an item by the extent it is physically able to produce a given effect, with no interest in, or reference to, the history or prior states of the system.

Some authors suggest that the two theories provide alternative definitions of functions that are conceptually independent; others suggest that the two theories are complementary and can be considered as special cases of a unitary definition of function (reviewed in Mossio et al. 2009). All these approaches focus on the analysis of the concept of function and its theoretical definition, i.e. mainly referring to what object(s) the term denotes in the world.

As Mitchell emphasizes following Cartwright (1986) these disputes can be better understood by spelling out what explanatory enterprises are making appeal to functional ascriptions. My working hypothesis, following the same strategy, is that the two theories "are different in that they are designed to analyse the explanatory ability of function claims as answers to two different questions" (Mitchell 2003, p. 112).

The Dispositional Theory fits well with the common approach in molecular biology, where it is common to use functional definitions of explanatory parts. The Etiological Theory, instead, is encountered in evolutionary arguments, where function is related to what is naturally selected for during an evolutionary process. *The epistemological integration of the two notions* of function and explanatory theories might be not only possible, but even necessary when looking at the kind of dynamics we have seen in the process of carcinogenesis. The explanatory structure of functional accounts can be, in fact, understood in its *dual characterization* that, in scientific practice, has its counterparts in the molecular explanatory models and in the systemic ones. Which of them is more able to perform the most comprehensive explanation of dys-function in biological systems follows what I have presented since Chap. 4. At this stage, what I would argue is that a more articulated reflection

---

[1] This theory is often, and especially in more recent literature on function, related with Wright's work (e.g. Wright 1973). However, as Jim Lennox pointed out to me, Larry Wright's theory of function cannot be reduced to the etiological account. Wright's discussion of teleological explanations, actually, seems to be particularly promising to further understand the use of functional explanations in science as well.

[2] This theory of function is usually related to Cummins (1975). Although other terms are used to define this theory (Systemic or Causal are the most common ones) we will mainly refer to them in their original form or we will use them as equivalent to 'dispositional'. Referring to this account in dispositional terms makes more evident, in fact, the epistemological approach to the notion of function that characterizes it. On the contrary, for example, the term "Causal Role" is confusing because a causal dimension is assumed by both theories. For a wider discussion on this point and related literature cf. Bertolaso 2011a.

on the notion of system and organization is required in theories of functional accounts. The most challenging feature of complex dys-functions is, in fact, that they ultimately affect *both* organizational levels and parts of an organic system.[3] Accordingly, we see that both theories of function require reference to a normative and teleological dimension, so that their heuristics might eventually converge when analyzed and applied in the explanantory models of carcinogenesis.

A new conceptual framework was recently proposed to solve the teleological and normative issues related with the notion of function, while bringing into a unitary picture the perspectives adopted by the etiological and dispositional theories (Mossio et al. 2009). This framework is known as the *organizational account* of functions.

In the organizational account, functions are inherently related to the idea of self-maintenance of the biological system. The teleological and normative dimensions of functions in this organization of the systems can be acknowledged by focusing on two of their fundamental properties, crucially involved in the grounding of functional attributions: *organizational closure* and *organizational differentiation*. These properties justify explaining the existence of a process by referring to its effects, so that "a process is subject to closure in a self-maintaining system when it contributes to the maintenance of some of the conditions required for its own existence" (Mossio et al. 2009, p. 13). Moreover, because of the organizational closure, "the activity of the system has an intrinsic relevance for the system itself, to the extent that its very existence depends on the effects of its own activity" (*ibidem*). According to the organizational account, the definition of a function implies the fulfilment of three different conditions. A trait *T* has a function if and only if:

- C1: *T* contributes to the maintenance of the organization *O* of *S*;
- C2: *T* is produced and maintained under some constraints exerted by *O*;
- C3: *S* is organizationally differentiated (Mossio et al. 2009, 16)

Interestingly the authors observe that dysfunctions take place when a *T* meets C2 and C3 but not C1. Now cancer is a disease related to an aberrant differentiation of cells and general dis-organization of tissues and organs.

7.2.2 Functional Assignments and the Inescapable Teleological Dimension of Functions

Reductionist models of the neoplastic process strived to attribute functions to some parts that could explain the general behaviour of the studied phenomenon. In the SMT, there are different molecular parts that, organized in a circuit, account for the functional properties of the cell and its behaviour. The system properties and behaviour are claimed to be fully explained in terms of the properties of the parts and their

[3] This also meets Artiga's concerns regarding the limits of organization accounts of functions when addressing intra- and cross-generation functions (Artiga 2011) and interesting questions already addressed in previous studies regarding the normativity of functional accounts (Saborido 2012).

mechanical interactions, and properties of the parts have to be considered intrinsically autonomous from the system. The fundamental effort of functional assignments in cancer research is thus related to the goal of making them independent from the teleological element that intuitively characterizes them. However, it is easy to show that despite all efforts reductionist models eventually require *normative criteria* to hold the explanatory potential of functional ascriptions.[4]

As we have seen in the previous chapters, the broad consensus obtained by the Multistep Model formalized by reductionist supporters (Vogelstein and Kinzler 2004, see 2.5) was due to the fact that it seemed to provide a scientific basis to the concepts of initiation, promotion, progression and transformation, which, until then, had been, essentially, only observational concepts. The assumption was that both the genesis and progression of cancer are mediated by a sequence of several molecular causes ultimately responsible for its occurrence (Hanahan and Weinberg 2000). Consistent with this, the employed terminology continued to be of tumour *progression* (rather than of the neoplastic process or tumour development) and the tendency was to treat cancer as an identifiable and circumscribed entity.

The effort to account for specific tumour functionalities through molecular parts, however, just multiplied the number of genes apparently linked to neoplastic initiation and progression. Therefore more and more complicated models started gathering more elements from various levels of cellular organization, which could be able to account, at some stage, for the final onset of cancer. The picture started to appear circular, through the identification of patterns and pathways all of them eventually interconnected (Vogelstein and Kinzler 2004), up to a "synthetic" image like the circuitry of the cell described by Hanahan and Weinberg, forming an almost unlimited combinatorial system of heterogeneous elements (Hanahan and Weinberg 2000). Overall, the effects that had been considered critical in the identification and definition of the molecular parts involved (proliferation, in vitro transformation) turned out to be largely coincidental to the tumour phenotype, which was ultimately identifiable through its metastatic capacity, and its ability to form tumours in other organisms.[5]

From another point of view, we can say: the cell becomes the context in which molecular pieces play their functional role. What is relevant to our argument here is that to make sense of the major properties justifying cancer behaviour – proliferation, metastasis, invasiveness – we need to refer to a cell, an *embodied biological structure* for which temporality is intrinsically meaningful (i.e. implies a natural history), and not simply to the genes, to which we initially attributed these properties. The explanatory framework, however, is the same. Moving within a reductionist

[4] I am not distinguishing here between normative and teleological accounts of functions. A useful analysis has been done by K. Neander. She argues that these two notions actually coincide in the biological field (cf. Neander 2009 and Bertolaso 2011a, b for further discussion of this point).

[5] As with Nagel's eliminativism, reductionism seems here to be facing a fatal objection, because it ignores the principal difference between a functional effect and an accidental effect, where accidental does not mean simple occasional effect. In both cases the effect can be constant and typical, but it is accidental in relation to functional effects that specify and characterize the phenomenon as a whole.

perspective, we have reconstructed a part, defined through functional ascriptions entangled in a much more complex biological structure: a useful tool to describe a sequence of events through hierarchically organized parts.

The difficulties related to the evidence that a seemingly unlimited number of molecular factors are involved during stochastic cancer evolution (making it almost impossible to establish a unique causative relationship among specific gene mutations/epigenetic alterations and cancer progression), were overcome by fixing the initial conditions at higher levels of biological complexity, through the concept of CSCs. The Hierarchical Model of cancer (Sect. 2.7) emerged as a better explanatory system to account for cancer heterogeneity and temporal dynamics. Under the Hierarchical Model, only a small subpopulation of tumour cells can proliferate extensively and sustain the growth and progression of a neoplastic clone: an almost linear sequence of events involving the new functional unit that is the cell itself. The normative dimension of functional ascription is delegated to the potentiality to generate offspring with the same behavioural pattern, i.e. uncontrolled, although to some extent differential, proliferation.

CSCs, by definition, carry all the necessary functional features to account for the final properties of cancer. Proliferation belongs to a cell, not to a gene, pluripotency and differentiation belong to a tissue-based context and not to a single cell, etc. In fact, in experimental practice comparison between functionalities is required as a final test for the neoplastic character of CSCs. Functional statements can be translated into ordinary causal statements with no loss of content only when they are recognized *at the level at which* functional properties can actually be observed.

Problems arise again when considering that the behavioural features that are relevant to cancer do not, in fact, belong to the neoplastic phenotype of cancer cells, but to the biological concept of *stemness* that characterizes the previous physiological identity of the biological units: stemness is a context dependent property. Differentiation and pluri-potentiality, by definition, biologically imply the influence of factors outside the cell, such as growth factors, tissue structure, and physical forces. Defining the explanatory system in cellular terms is not enough to account for those processes, also because molecular parts present different functional features, depending on the different steps of the developmental process. A relational dimension reappears here as an intrinsic feature of the definition of the system itself. Despite the advantage of the hierarchical account of CSC origin, the explanatory strategy adopted by this approach still seriously underspecifies functional ascriptions, which, in turn, generate an epistemological problem (Gupta et al. 2009).

Hence, problems related to the infinite regression of the attribution of functions are just apparently overcome. The huge debate still going on in the literature about the CSC concept reflects, in my opinion, this limit. The insights gained with the Hierarchical Model of cancer are lost when the contextual environment (i.e. the organism and its structural-functional organization) is obliterated again. Although crucial functional ascriptions do not even belong to genes alone, but to cells, functional statements are still formulated in terms of a part that is able, in its own right, to produce a given effect within a wider system.

While the reductionist account assumes parts with a static nature, systemic perspectives are better equipped with the epistemological tools to discuss the organizational level as the object of inquiry. This is why they usually stress the relevance of the context. It is not enough, in fact, to say what a given trait is actually able to do in given circumstances; it is necessary to put forward adequate assumptions that hold what that trait is supposed to do. The involvement of the functional trait, in the explanatory accounts of dysfunction, requires this integration. Functional attribution thus assumes a normative dimension to account for what that trait is supposed to do.

A reductionist-mechanistic epistemology is useful and practical for analysis of *short-term* changes in linear systems that are not subject to rapid environmental perturbation, i.e. for analysis, where the *temporal dimension* is not relevant. However, if cancer is not a thing but a process, such temporal dimension should be taken into account at some point. The regulation of the gene product, and hence its functional activity in cancer, may occur at various stages of the maturation of a protein and is usually mediated by time-dependent physical and chemical contextual factors, as it happens in other contexts of functional biology. The peculiar timing of organic processes is always involved in all levels of the biological organization. The behavior (B) of the parts is also a function of time (t), but such timing does not belong to the intrinsic behavior of parts in isolation, nor to their mere aggregation. Several examples are described in the literature (Gilbert 2005). This implies that $B(t)$ does not follow ergodic approaches but requires systemic ones.

Although at some levels the properties of parts may effectively play a causal role to some dynamics, such role is explanatorily relevant for higher-level properties, when the phenomenon is analysed as a whole (usually reflected in the relative terms of the explanatory account: see also Sect. 7.2). In an organism, it is not enough *to tell a cell what to do*; it is necessary to clarify *what is to be done at one given time and not in another* through environmental signals. Such signalling context is what I call a 'microhabitat'. In the physiology of the organism the signalling role is played by hormones. The biological properties of cells in an organism have thus *to be space and time dependent*, as in the case of stem cells, because of the historicity of the body, i.e., its *developmental ontogeny* that constitutes a *real obstacle to the success of reductionism* (Soto and Sonnenschein 2005; Soto and Sonnenschein 2006b).

Reductionism relies upon units whose identity is logical, not bio-logical, i.e. follows the logic of what is necessary, not of what is relational. The epistemological price paid for reducing the teleological aspect, implied in the concept of cellular differentiation, to a simple derivation of parts, is the actual exclusion of a teleological dimension as a proper explanatory aspect, triggering a search for another satisfactory foundation of the normative dimension. When recovering this perspective through the notion of natural selection, (i.e. by reference to external biological laws rather than to purely molecular ones), the dismissed teleological dimension represents itself anew. The historical dimension, typical of the etiological theories of function, reappears through an explicit reference to evolutionary principles.

7.2.3 The Limits of the Evolutionary Argument and Selected Functions

Biological functions intrinsically appear as means–end relationships. In other words, biological function presupposes a teleological-normative dimension to the extent that it refers to some effect that the trait *is supposed to* produce. As we have seen, in fact, the Cell-Centred Perspective employs active terms that describe tumour cells, such as 'program', that culminate in the metaphor of the 'renegade cell' (Weinberg 1998).

Within the reductionist paradigm, the recognition of a causal role is founded on temporal priority. For example, the causal role of the epigenetic component is founded on the temporal priority of the rise of epigenetic over genetic alterations.[6]

The Stochastic Model and the Evolutionary Argument reconstruct a system of logical coherence in which an (efficient) causal relation is assumed as being explan-

[6] Feinberg provides us with some considerations clarifying the implication of philosophical reductionism: "Epigenetic changes can provide mechanistic unity to understanding cancer" (Feinberg et al. 2006, p. 25, see also the Appendix for more context). The multistep model, that had put the molecular elements into a sequence needed, in fact, to integrate the schema of neoplastic onset, by means of a polyclonal model in which, not only the genetic, but also the epigenetic gene expression regulatory components, in the broad sense of the term, have to be considered. The Epigenetic Model seemed to have given an answer to the problems of cancer latency and heterogeneity that previous models could not explain. Therefore, as per an already mentioned quote, almost a guarantee for the validity of the model and its legitimacy within the reductionist paradigm, the recognition of a causal role in the epigenetic component is founded on the temporal priority of the rise of epigenetic over genetic alterations. How can an argument for causal argumentation of epigenetics changes be supported? The establishment of a causative relationship has been always molecular biology's goal and passion, but the argument for causality has always been posed in terms of temporal priority. So that a convincing causal argument is made mainly through different arguments, like the evidence that constitutional epigenetic alterations are linked to cancer risk, as it has been demonstrated in Beckwith–Wiedemann syndrome and Wilms' tumour, etc. Thus, "the epigenetic change precedes cancer and confers risk for cancer, a strong argument for causality" (Feinberg 2007, p. 437).

atory of a phenotype such as that of cancer. This causal relation is however not applied to the molecular parts, but to an external principle, natural selection, when the Cell-Centred Perspective has to account for change over time of the tumour cell phenotype. Natural selection plays the deterministic role previously delegated to genes. The functions that define a neoplasm, following a deterministic logic of this type, are structured by means of chance and natural selection that acts as an external law, stabilizing the inter-actions between the parts, causally organized in a linear hierarchical system. This concept of function reduction allows reductionist models not to deal further with the origin and nature of the observed capabilities, and implement a simplification of the biological complexity by means of the identification of a system that deterministically controls this very complexity. Functions don't need to be explained, they become the explanation.

An evolutionary explanation of cancer features assumes that the development of the disease is linked to the progressive accumulation of mutations in tumour suppressor genes and oncogenes through progressive selection of more malignant cells. It is thus commonly accepted that "The development of cancer is an evolutionary process that is driven by multiple genetic and epigenetic changes" (Nature Editorial 2009). In this way, most, if not all, cancers *have acquired the same set* of functional capabilities during their development. The properties of cancer are then defined by restricting functional attributions to those (higher-level) capacities constituting the 'goal states' of the system: self-sufficiency in growth signals, insensitivity to antigrowth signals, capability to evade apoptosis, a limitless replicative potential, sustained angiogenesis, capability to invade and metastasize (Hanahan and Weinberg 2000).

The evolutionary argument gained wide consensus, complementing dispositional functions with an etiological story. The causal role of gene function was integrated by evolutionary functional arguments to account for the (dys)function of the traits, in this case, the genes (Hanahan and Weinberg 2000). The dominant feature of genes and tumour cells in their proliferative activity, clonal expansion and metastatic final capability, was explained in a perspective similar to the 'selfish gene' (Dawkins 1976). Natural selection embodies once again a normative dimension, one which is almost intentional in feature, and the concept of biological function is reduced to the one of *efficient causal relation*: the identified systems are "capable of", machines that, once the specific properties and conditions have been defined, will perform certain functions. In philosophy, this position is referred to as 'propensity view', identifying functions with *causal contributions* of components to the life chances, or fitness, of the system (Heng et al. 2009; Bigelow and Pargetter 1987).

The evolutionary account of cancer, however, underwent major criticisms. A selective advantage of the phenomenon of cancer, or the metastatic properties of its cells, is difficult to reconcile with the biological logic of organism-based selective advantage. The evolutionary "goal contribution approach" (Mossio et al. 2009),[7] much like the 'dispositional account', is coherent with the idea that any trait should change its specific contribution according to the particular conditions, and each trait will end up possessing an indefinite list of actual functions. In a dispositional view, then, we lack the theoretical resources to distinguish between functions and accidental contributions to a goal state. More than by an *ad hoc* and *a priori* universalistic evolutionary argument, then, metastatic properties seem to be explainable by studying the previously existing healthy cells (Germain 2012).

A more radical problem is the inherent *local* character of natural selection. A hypothetical natural selection process among clones should be *strictly dependent on the current cellular environment*. We may then encounter difficulties when *metastasis* has to be explained through supposed 'metastatic properties' acquired in a local cellular microenvironment. To conform to such locality, evolutionary models of cancer articulate evolutionary theory in terms of fitness (cells proliferation), as a 'real adaptation' – which would go beyond the local immediate context – but this seems hard to state for cancer cells. In general, the terminological uncoupling of fitness and adaptation doesn't solve the problem. As McLaughlin observed, the distinction between having a function and being an adaptation is just an attempt to show the "metaphysical innocence" of functional notions by reducing the former to the latter and basically, giving up with their normative dimension, eliminating them (McLauglin 2001). We must admit that almost all biological activities can be taken to the level of function within any system to be explained.

In my view, at least in the context of cancer, stochasticity and evolution seem to be more adequate to explain survival events, than to enhance the appearance of new stable phenotypes and the transmission of a stable genomic pattern showing a real adaptive behavior. If carcinogenesis is the *explanandum*, any reference to evolutionary theory to identify goal-directedness seems to be an *ad hoc* addition to an explanatory theory unable to properly define functions, nor to distinguish them from dysfunctions. This context makes cancer cells appear as physiologically healthier than normal cells. Normativity of functions needs then to be based in a different field. Otherwise a principled criterion to identify the relevant set of contributions,

[7] By 'goal contribution approach' (GCA) (Mossio et al. 2009) is meant an approach that links the concept of function to the idea of goal-directedness, introducing more specific constraints on what makes causal relations properly functional. To identify the goal states of a system following a naturalized, and non-arbitrary criterion, the GCA has adopted a characterization of goal-directedness where biological systems can be described as having as their essential goal to survive and reproduce. Hence, biological parts are dispositions that contribute to these goals: still within a Dispositional Theory of function. In the case of malignancy the goal might also be defined in terms of survival advantage. However, even assuming a goal of the system, Dispositional Theories – though providing a general characterization of 'useful' contributions – are not able to distinguish between proper functions and accidental, or secondary, contributions.

for which functional analysis makes sense, is still missing and the circularity of functional explanation is not overcome.

7.2.4 Harmonizing Views of Function

The Dynamic and Relational View allows for an integration of the two theories of function through a wider concept of causal notion which includes causation 'by holding' (Sect. 5.2.4), and through a more dynamic view of scientific explanation (Chap. 6). McLaughlin (2001) claimed that the most genuine functional explanations involve not so much an illicit explicit appeal to final cause, as an implicit appeal to *holistic causality*, which is in fact recovered by any interpretative model of cancer (reductionist and systemic), through an 'organizational' notion (although, obviously, used with a different meaning due to the different epistemological frameworks). The Dynamic and Relational View, with the idea of Operational Integrative System, refers to properties and causal notions that belong to the Operational Integrative System as a whole, and are not reducible to the sum of parts, however identified and defined. Self-organization through parts differentiation is clearly a critical feature of complex biological systems and of their most articulated regulatory pathways. *Once* the object of inquiry is defined through mesoscopic style of reasoning, dispositional accounts provide the most typical view of functional attribution in ordinary empirical research (systemic models, the TOFT and other organicist perspectives go back, in practice, to a dispositional account of function).

In the Dynamic and Relational View, functions and dysfunctions are no longer related with the presence, or absence, of a trait but with the (functional) integration of parts in the mesosystem (Sect. 6.3) that, in turn, accounts for their functional features. This change marks a relevant switch in the explanatory perspective creating a rupture with the traditional models of cancer. Functions are analysed from the point of view of the organizational enterprise of the whole system and not of the traits they (seem to) belong to. In these models, the identification of the mesosystem which generates a particular (physio)pathological phenomenon, becomes a key prerequisite for the explanatory enterprise. The identification of the system is a priority for the study of (dys)function and the size of the case is taken as an intrinsic component of the system, not 'explained' by the model, but rather serving as part of the explanation.

The identification of the level of analysis of a physio-pathological phenomenon is a key point of the whole explanatory enterprise. The Operational Integrative System (the organism) is considered as a whole that influences, and also determines, the properties of its parts. The reference system cannot be a cellular system allegedly generating a hierarchical structure characterized by phenotypic heterogeneity, as claimed in the Cell-Centred Perspective; it is the constituted organization of an organism itself that makes sense of the phenotypic hierarchical heterogeneity of cancer, once the pathological conditions for carcinogenesis are settled. The historical dimension is always recognized as a fundamental dimension of the biological phenomena to be studied.

As we have already seen above, most of the problems related to functional attribution within the reductionist perspective concern the relevance of the (micro) environment in the neoplastic process. The ultimate demonstration that a modification is carcinogenic is the ability to show malignancy *in vivo*, i.e. the ability, upon transplantation, to produce tumours (Sect. 6.4.1). This demonstration is further complicated by a logical consequence of reductionist assumptions. In fact, the inability of a transplanted cell to cause malignancy does not necessarily mean that it is not tumorigenic; it could just be that it was not transplanted into the proper context.

The point is that neither single molecules nor structured elements alone are adequate to capture the specificity of biological regulatory processes and, thus, to explain the phenomenon of cancer in all its phenotypic characteristics. What is required instead is always *a comparison of contextually identified functional features*. This also makes sense of the observation that a tumour cell can be reverted into a non-malignant cell if placed in the right environmental context.

7.3 Making Sense of Problems with CSCs

Now let us attempt an application of the emerging relational theory of explanation to make sense of the paradoxes and difficulties of one of the most advanced Cell-Centred models in the history of cancer research: the Hierarchical Model (Sect. 2.7) with the related concept of Cancer Stem Cell (CSC), integrated by epigenetics (Sect. 2.6) and by the evolutionary arguments (Sect. 2.8). This is just an example to demonstrate that the Dynamic and Relational View has tight links with scientific practice, and may help in improving understanding of experimental studies and interpreting results in cancer research.

As discussed in Chap. 2, the Somatic Mutation Theory assumed that neoplastic properties could be acquired by means of genetic mutations, clonal proliferation and selection of tumoral phenotypes only. Later on, epigenetic changes were found to be causally relevant for the origin and stabilization of the neoplastic phenotype, also regulating the expression of many Tumor Suppressor Genes (Greger et al. 1989). Epigenetic changes could be considered either alternatives, or surrogates, of genetic alterations (Egger et al. 2004) on GK (gatekeeper) genes, as in the case of the inactivation of TSG, or on CT (caretaker) genes, as in the case of the activation of certain ONG (cf. Sects. 2.5 and 5.2.2). With particular reference to the early steps of oncogenesis, initiation and progression, the Epigenetic Model came to alleviate the contradiction between epidemiological data and the statistical data calculating the time required for a mutation to be stabilized by natural selection (Feinberg et al. 2006). The Epigenetic Model could also integrate the genetic perspective with a regulatory one through epigenetics and thus explain the heterogeneity of tumoral phenotypes through the epigenetic alterations in stem cells (Feinberg et al. 2006).

Feinberg and colleagues suggested that:

> …epigenetic disruption of progenitor cells is a key determinant not only of cancer risk, but of tumour progression and heterogeneity late in the course of the tumours that arise from these cells. Epigenetic changes can provide mechanistic unity [my emphasis] to understanding cancer, they can occur earlier and set the stage for genetic alterations, and have been linked to the pluripotent precursor cells from which cancers arise. Importantly, early epigenetic changes could explain many of the heterogeneous properties that are commonly associated with tumour cell-growth, invasion, metastasis and resistance to therapy. To integrate the idea of these early epigenetic events, we propose that cancer arises in three steps: an epigenetic disruption of progenitor cells, an initiating mutation, and genetic and epigenetic plasticity (Feinberg et al. 2006).

The Epigenetic Model joined into the CSC model (Sect. 2.7). As it has been explained extensively in Chap. 2, the fact that epigenetic changes can be present in the first steps of cancer genesis, or even in normal tissue, before the appearance of a tumor, was taken to suggest that early epigenetic changes in stem cells could contribute to a unified view of cancer etiology.

Interpreted in our Dynamic and Relational View, the development of the original clonal model into an epigenetic one appears as an endeavor to make sense of some *context dependent features and of the causal relevance of epigenetic factors within the microenvironment.*

According to this model, cancer arises in three steps: after an epigenetic alteration of stem/progenitor cells within a given tissue, a mutation of genes – among those known as relevant in cancer origin and progression – follows. Finally, genetic and epigenetic instability arises, which eventually leads to increased tumor evolution. Feinberg, Ohlsson, and Henikoff say, "Note that many of the properties of advanced tumors (invasion, metastasis and drug resistance) are inherent properties of the progenitor cells that give rise to the primary tumor and do not require other mutations (highlighting the importance of epigenetic factors in tumor progression)" (Feinberg et al. 2006). The relevance of the epigenetic perspective assumed in this model is also highlighted by an assumption made by these authors: alterations in the stem/progenitor cell "can be due to events within the stem cells themselves, the influence of the stromal compartment, or environmental damage or injury" (*ibidem*). Moving from the clonal to the Epigenetic Model was not just a matter of a more complicated model: the level of explanation changed; the chosen causal model system actually changed the concept of Progenitor Cell and the *explananda* narrowed down to the different tumor properties of these Progenitor Cells.

7.3.1 Explanatory Overload on CSCs

The CSC Model accomplishes a partial reduction in Schaffner's terms (Sect. 6.3.2). The general evidence that cells are involved in the development of cancer, that epigenetic changes are relevant in this process, and that the timing of the process itself is not (only) driven by genetic factors, are captured by explaining these features in terms of progenitor cells and by widening the characterization of these cells through the notion of stem cells. By assuming that something relevant is entailed by the notion of stem cells, it is possible to fit parts and events into a linear causal sequence. The Preferred Causal Model System (PCMS) is chosen and it works in explaining the polyclonal origin of cancer. The relevance of one causal factor with respect to another is justified in terms of the temporal priority (Feinberg 2007) so that the PCMS captures the temporal dimension implied by a mechanistic account of biological processes. The explanatory scheme that emerges is a serial sequence of mechanically described events. The PCMS enacts a context of effects, a peculiarity that recalls the biological concepts used in systemic models of carcinogenesis.

The inter-level character of PCMS is also clearly exemplified here, showing cells, parts of cells (genes), regulatory devices (epigenetic alterations), as well as connections among them (clonal derivation), ultimately resulting in the polyclonal feature that characterizes tumor masses. The inter-level character of the model is not only characterized by the involvement of genes, epigenetic factors, and cells, but it is also grasped by the additional condition of stemness of those cells, i.e. the specific behaviour of those cells seen as parts of a larger entity. As in other previous examples (Schaffner 2006), the model is simplified and idealized, and uses causal statements such as "results in different rates of cancer spreading".

The Hierarchical and epigenetic Model couples the timing sequence of neoplastic progression with cell differentiation and tumor heterogeneity by hypothesizing that only a small subpopulation of tumor cells can proliferate extensively and sustain the growth and progression of a neoplastic clone. A (dis)organization of differentiated units is an explanatory system that accounts for cancer heterogeneity and its temporal dynamics, i.e. different stages of differentiation, some of which retain the tumorigenic properties.

The PCMS identified through the notion of CSC is thus expected to explain the different rates at which cancer cells produce new cancers and cellular aberrant differentiation. But the production of differentiated *non*-tumorigenic offspring by those cells is, logically speaking, contradictory, given the properties of a stem cell.

The stemness of progenitor cancer cells is relevant to the adequateness of the mechanic explanation for how not all tumor cells retain neoplastic properties. The stemness character of progenitor cells allows a hierarchical structure of cell differentiation. The stemness of progenitor cancer cells is thus a qualitative feature of the *explanans* for this phenomenon, but *it hardly accounts for a next-step-level process: cell differentiation.* In fact, the tumorigenic character of the progenitor cell might imply bidirectional interconvertibility between CSCs and non-CSCs. *The assumption is that tumour and stemness properties belong to the same cell in the same way.* The conclusion reached by some authors, therefore, gets back to previous biological definitions of stem cells, so that CSC representation may be a function of the cell type of origin, stromal microenvironment, accumulated somatic mutations and stage of malignant progression reached by a tumor. Accordingly, the CSC model must stand or fall on the basis of punctual experimental characterizations of cancer cell populations (Kelly et al. 2007; Quintana et al. 2008). The priority of the linear, although branched, explanatory sequence of events over the biological (functional) properties of the parts, and of them as a whole, makes it difficult to go forward through functional explanatory accounts, i.e. attributing new functions to the same parts in order to account for the overall behavior of the systems they belong to.

7.3.2 From Hierarchical Descent to Dynamic Regulation

The generalization of the concept of stem cell created a tension (or even incompatibility) with its *molecular* identification. In our theory of explanation (Chap. 6) this is natural and coherent, since the *explanandum* – in this case, tumor heterogeneity – is not neutral with respect to the identification of the mechanisms that are explored afterwards or to the identification of the level at which the phenomenon needs to be explained.[8] When treating the biological units in isolation, we are pushed to question again the actual existence of these units and their explanatory potential in all solid kinds of tumors (Gupta et al. 2009).

If, in one model, CSCs "produce aberrant differentiation" or "result in an aberrant hierarchical organization of cells in tumors", nonetheless *the biological concept of stemness is heavily determined by the cells' position in the tissues.* Differentiation and pluripotentiality depend biologically on growth factors, tissue structure, and physical forces, that by definition do not belong to the cellular level. They define the state of parts that may explain some features of the biological process. The context needs to be highly controlled in order to promote the specificity of

[8] Somebody might ask why heterogeneity isn't explained by multiple mechanisms. A simple answer could be that what mechanisms would eventually explain is: this kind of heterogeneity still depends on the PCMS. Different examples could be shown from the literature (e.g. stochastic models of cancer, the hierarchical model, a population model, etc.).

the program itself. In our Dynamic and Relational View, a complete objectification of the system is not possible while: “The study of CSC biology is predicated on the ability to accurately assess CSC representation within cancer cell populations” (Gupta et al. 2009, p. 1011).

The role of the context is not just a generic epistemological assumption for mechanisms to work, but its specific features – that are different depending on the PCMS – do play a role in the adequacy of those mechanisms as well. Although some dynamic properties can be adequately captured in mechanistic terms, other features are relevant for the whole explanatory structure of the model and have nomological relevance. Hence, the insights gained with the CSC and Hierarchical Model of cancer are lost when the overall structural-functional organization of the organism and its relevance in framing the mechanistic account is overlooked.

From a complementary perspective, we could say that inter-level regulatory properties require any mechanism to be framed in the adequate PCMS, and that they don’t concede any self-referential explanatory power to the mechanism.

Let’s tackle the same issue from the point of view of “function” (cf. 7.2). *Just because something actually works and has effects that does not mean it has a function.* As in biological processes the system itself keeps changing, the normativity assumed also in physiology, through dispositional account of functions, implies not just physical derivation of parts from an original one, as guaranteed in the Hierarchical Model, but also that functions themselves have to date back to that original part, both in explanatory terms and in their own definition. In other words, if this model apparently recovers a specific dimension of etiological or evolutionary theories of functions, the reductionist perspective is not able to hold that explanatory dimension from inside. Sticking to the dispositional or causal role theory of function, the Hierarchical Model fails to draw a principled demarcation between systems whose parts appear to have, or not to have, functions (Moreno et al.).

The problem is that the hierarchical dimension of the “Hierarchical Model” is just related to a mechanical aspect: *derivation* expressed in terms of proliferation. By definition, the elements are expected to guarantee inheritance of a neoplastic phenotype to their progeny and to admit heterogeneous differentiation at the same time. Moreover, both capabilities should be explained only in terms of CSCs properties, according to the theoretical premises that led to the identification of this cellular system: their capacity to contribute to the emergent properties of tumoural clones should be exercised whatever the context and the function of its own biological regulatory system.

In summary, the concept of CSCs does not seem to be able to fully explain all the aspects that also characterise the dynamic process of carcinogenesis. A full explanation is only possible when taking into account their functional context, which is a combination of both hierarchical organization and dynamic processes.

This case study helps us understand better the problems affecting simple mechanistic reduction. It also shows how the PCMS might be a solution, through its relationship with the formal conditions for partial reductions proposed by Schaffner and a defense of their non-reductionist character. Because of their mechanistic structure, these explanations have to consider *the difference between the epistemological role of the parts and the system behaviour*. This distinction resolves the apparent circularity and ambiguity and also explains why we had to develop a new theory of explanation where inter-level regulation plays a crucial role in defining the *explanans* itself.

Chapter 8
Conclusion: Beyond Dichotomies

Cancer is a very interesting case study because it is a process deeply connected with the relational principles that not only structure, but also maintain, the functional and structural organization of living systems in time and hold our capability of grasping them through scientific practice. In fact, research programs move towards the development of approaches and conceptual tools able to explain how biological systems develop and what they do (or fail to do) in relational terms.

Building upon the deeper implications of systemic approaches and anti-reductionist claims, I have proposed a Dynamic and Relational View able to make sense of the compatibility and reciprocal role of different strategies and points of view that have emerged in cancer research. A deeper attention to the process by which scientists discover what features and dynamics are relevant in explanatory terms of a given phenomena is necessary in order to understand cancer and other multi-level regulatory processes. This process of 'understanding by relating' overcomes the dichotomy between reductionist and anti-reductionist accounts in the scientific literature. The relational account of biological interactions closes the circle.

The same perspective on biological behaviours also justifies why *empowerment* seems to be the right and best way to improve the capability of a biological entity. The growth of an organism can be aided by strengthening the viability conditions of its functionalities so that the overall behaviour can take place with a higher degree of systemic coherence and possibly at a lower level of energy. It is commonly recognized that, in the end, the best cure for cancer is prevention. But prevention is nothing but working in terms of empowerment of the system. It is possible because any biological relationship is embodied in a context or microhabitat that is made of belonging networks. This way of understanding cancer cells' behaviour requires, at some point, acknowledging dependencies more than autonomies. What is compromised in the neoplastic process is not a function of a cell or a gene in a cell, but the conditions of possibilities of functions that are no longer constrained. The tumoural phenotype appears rigid, not plastic, in this framework, and this fits well with evi-

M. Bertolaso, *Philosophy of Cancer*, History, Philosophy and Theory of the Life Sciences 18, DOI 10.1007/978-94-024-0865-2_8

dence that the neoplastic phenotype eventually requires lower level of energy than the normal ones. Compensation processes are thus the physical counterpart of the reflexive dimension of the causality I tried to describe, and empowerment takes place through coupling of dynamics and processes. By lowering the thresholds of energy reaction, the process of coupling creates at different levels of the biological organization new microhabitats, where new properties can emerge. Depending on the established dependencies, dynamics at lower levels can enter an automatic phase, as shown by the rhythm of the cell cycle also discussed by Sonnenschein and Soto since 1999. Therefore, as Waddington said "it is the understanding of the nature of the networks of interaction, which are involved in the process and which a collection of cells becomes organized into an organ with a unitary character, that still remain the central question when addressing living beings" (Waddington 1977).

We can now reconsider Sporn's dichotomies in cancer research (Sporn 2006, see the table in the Introduction).

Is the disease cancer or carcinogenesis? Is cancer a genetic or an epigenetic disease? Do we need to cure end-stage disease or to prevent early disease progression? These dichotomies can now be recomposed in the light of the epistemological priority of stability over specificity. We need a deeper comprehension of such stability and we need to be aware and committed to figuring out the adequate explanatory level of actual biological dynamisms through the identification of mesoscopic levels.

Is 'the whole greater than the sum of parts', and is this assumption opposed to reductionism? We now have the interpretative key of how the 'greater' within the expression 'the whole can be greater than the sum of its parts' should be understood. The limit of this expression is that it preserves the idea of 'parts'. There is a whole – a multi-unity dynamism – *beyond* the configuration of parts and wholes in time. There are biological functionalities that take the form of Operational Integrating Systems whose functioning unity demands to remain in focus beyond the analysis of constitutive parts. *Operational* stands for generative; *Integrative* indicates the morphostatic dimension. The interactions in Operational Integrative Systems imply more than mechanic feedback loops. They imply reflexive and synchronic features of the *dynamics that hold* the neoplastic process and recurrent reference to auto-stabilization and inter-dependencies.

Must cancer research be more hypothesis-driven or more observational? The 'observational' point of view emerges as relevant when relationships (not mere interactions but 'inter-actions') and regulatory processes (not mere control) are shown to be crucial in understanding biological action. Living beings are first and foremost systems of relational activity: they move in a specific way that is always relational in different senses (with respect to their intrinsic organisation, their integrative processes, their interactions with the environments and other living beings as well). Much more work can be done on the last dichotomy.

In the words of Sporn, any synthesis of the preceding dichotomies should concentrate on the genesis of cancer from its very onset: the scientific question about carcinogenesis. Moreover the idea of carcinogenesis being an *aberration* of the wound healing response constitutes a fruitful framework to observe carcinogenesis. Cell-Centred and Organism-Centred perspectives equally highlight this point that, with the relevance of temporal-contextual factors, constitutes the element of convergence between both positions. They differ in their understanding of how the context relevance and the aberration (of the wound healing response). In the Cell-Centred Perspective, contextual factors are juxtaposed elements in the causal story. In the Organism-Centred one, contextual factors are identified and understood in causal terms moving from the integrated response of the organic system. The final proper interpretation of both context relevance and aberrant differentiation comes from the considerations that (1) due to the "tremendous adaptive redundancy in the manner in which normal or premalignant cells respond to environmental stress" the ultimate way to deal with carcinogenesis is to suppress this disease in its earliest stages, before extensive DNA and protein damage occur (Sporn 2006) and (2) "We need new emphasis on epigenetics". Again the peculiar relationships that hold biological dynamics and regulatory issues push cancer research back to prevention or therapies that target "entire regulatory networks", i.e. dynamics in action. As I have highlighted, and also Sporn states, clinics point to a revision of reductionism. Cell-Centred models can be practical for analysis of short-term changes in linear systems that are not subject to rapid environmental perturbation. However, when dealing with long-range interactions and dynamics that are clearly time and context dependent we need systems-oriented approaches (as also pointed out by Weinberg in the paper quoted in the Introduction).

The debate between Cell-Centred and Organism-Centred models does not concern the compatibility/incompatibility between the two perspectives. It concerns their mutual implication when the *explanandum* is the origin of cancer. We went through this point at different stages of the volume, when discussing their explanatory independence and epistemological interdependence, the SMT as a specific case of the TOFT and when discussing the non-reductive dimension of explanatory accounts of inter-level regulatory processes. The operational feature of inter-relations can be seen from the point of view of the way of being of the system, i.e. its constitutive dynamics (Organism-Centred) or by analysing the dynamics in the system's activity (Cell-Centred). In the first case we get a clearer image of the different levels of the biological organization involved and the context is in the front line of the argument. Everything appears more dynamic and satisfactorily described. Different examples support this thesis: the definition of the tissue as a tumor suppressor factor, the expansion of the hallmarks of cancer in terms of enabling capabilities and the shift towards *viability conditions* instead of causes. In the Cell-Centred Perspective case, the focus is on the abstracted subject defined through external projections of the constitutive dynamics, and the habitat is mere context or environment. The tumour cells and the hallmarks of cancer are understood better from this perspective. We get a lot of details on the cell by explaining cancer (cf. the integrated circuit of the cell described in Chap. 2), but hardly understand its temporal dynamics, which ultimately are more relevant for clinical and therapeutic purposes.

This claim against genetic determinism does not exclude that genes do have – depending on the observed phenomena – an epistemological privileged status in a relational ontology and epistemology, but genes do not have "a privileged *metaphysical* status" nor are they "*the building units* of the organism" (Soto and Sonnenschein 2005, 104, my emphasis). Introducing causal factors that are not even of a molecular or genetic nature as in the case of the epigenetic models or of the attractor landscapes, means to appeal to processes of an order which is not only genetic and DNA-based, but regulatory and dynamic in nature. It also means to recognize that the higher-level properties in some way influence the functional configuration of the lower levels and that such influence comes in degrees.

A relational account is much more adequate to grasp biological dynamics and their scientific understanding. It can also conciliate a common dichotomy in biological sciences regarding the extent to which biological elements and parts can be considered autonomous or dependent. They are both. They have a relational nature that entails both. In this sense the expression of "the whole is more than the sum of the parts" can be understood. It is 'more' in the relational sense; it is new. And such novelty has a priority in explanatory terms when the scientific question regards biological growth and development.

Genetic or organismic or environmental factors are indeed causally relevant in cancer. The Dynamic and Relational View acknowledges such causal relevance (called 'specificity') putting it in the context of viability conditions (especially, 'stability' of the relational context). Genes are given back their explanatory role, with a new understanding of its nature and limits. We have more instruments to understand problematic notions such as 'recessive vs. dominant' functionalities.

A key concept I have proposed is "biological determinations". These are sets of regularities that emerge and hold, under particular conditions, the dynamisms that we can observe and study in the biological world. Biological determinations do not conflict with the deep indeterminism of biological entities: determination and indeterminism feed each other. They are faces of the same coin. Biological determinations are, at once, the material conditions of life and the prerequisite for the scientific study of life, which proceeds by their identification and by the choice of the appropriate 'mesosystem' that makes sense of them. In this book, cancer has brought us to appreciate the interplay between what – and how – we can get to know about biological entities, what – and how – we can explain, and the dynamic and relational picture of the biological world that is incredibly clear and elusive at the same time.

We live in a time in which personalized medicine is being proposed. In this context, we can read the Dynamic and Relational View of cancer as a plea for finding ways to balance the relationships, thereby creating the viability conditions for physiology, as opposed to avoiding or fighting pathologies. Scaling up to personalized medicine, the Dynamic and Relational View prevents us from interpreting 'personalized' as 'individualist': no living being is static or isolated; caring and clinical practices should create the conditions for patients to be less and less alone, more and more human. The same should be said for scientists and doctors and for the great role that science can have in humanizing the society.

Appendix: Clarifications on Reductionism and Mechanism

There is a huge body of reflections (and debates) behind many philosophy of science terms that are used throughout the book. Anytime talk about "reductionism" or "mechanism" or "emergence" it would be fair or perhaps mandatory to honour and discuss all distinctions and points of view that have been raised by philosophers and scientists, or at least those which are currently being discussed in the relevant communities. Take reductionism as an example. The philosophical discussion on the possibility of reductionism in the bio-medical sciences can analyse (a) the different dimensions of reductionism: ontological, epistemological, methodological; (b) the relationship between reductionist and anti-reductionist views in biological sciences (explanatory issues and laws); (c) the problems with reductionism in biology (like multiple realizability and context dependence). On the other hand, in this book I had an urgency to advance a positive proposal – the Dynamic and Relational View of Cancer – and I had to surrender to the fact that stopping and debating all the alternatives all the time would have been too confounding. So I opted for making my view as clear as I could in Chaps. 5 and 6, so that philosophers will be able to understand and criticize my arguments directly. At the same time, I decided to write this Appendix, to save some of the complexity of the existing debate (also for non-philosophers) and to provide philosophers with the due coordinates to place my work.

Dimensions of Reductionism

Speaking of reductionism, the debate in scientific literature on the different dimensions of reductionism have historically been articulated in the following way (Ayala 1974; Ayala and Arp 2010):

- *Methodological reductionism* is mainly related to the strategies for research and acquisition of knowledge in empirical sciences. Science often proceeds through analysis, and – when explanation is the goal – resorts to the lowest possible level.

M. Bertolaso, *Philosophy of Cancer*, History, Philosophy and Theory of the Life Sciences 18, DOI 10.1007/978-94-024-0865-2

Methodological reductionism is commonly accepted by both scientists and philosophers as a legitimate approach, but there is discussion about what the "lowest possible level of explanation" means, and about the epistemological status of an analytical approach. The latter, in fact, taken only in an ontological and metaphysical perspective, would mean that molecules, as the smallest physical parts of biological entities, are the ultimate valuable terms of analysis and explanation. Within this perspective, biologically relevant explanations will be obtained only by investigating physico-chemical processes because of their explanatory value. Against this, anti-reductionist positions would claim that these processes do not belong to biology and are not relevant at all.[1]

- Historically, *epistemological reduction* was mainly related to *inter-theoretical reduction*, so that it would be possible to replace one theory *T* with another one *T'* which is more 'explanatorily' powerful. Theories and experimental laws get reduced when they turn out to be special cases of theories and laws formulated in some other branch of science. In the last decades, however, there has been a shift from discussion about theories to discussions related with explanations and the role that mechanisms, emergent properties and biological concepts play in explaining biological phenomena. The latter approach has been often adopted in volumes published on the possibility of reductionism in the biomedical sciences. A classical issue is whether biology can be reduced to physics (and chemistry, as many often add). Nobody would currently define himself a *vitalist* who claims that living processes are, at least in part, the effect of a non material entity (vital force) (Ayala 1974, viii). Under the common background assumption of *materialism*, however, epistemological reductionism poses the question whether one theory about the world will be plausible and achieved once the world structure, through the structures and relationships among sciences, is clarified.
- *Ontological reductionism* is mainly related with the question about what exists, and about whether physical, chemical entities and processes underlie all living phenomena. There are different kinds of materialism, so that ontological statements about life conceived as a highly complex pattern *of* physical and chemical processes need to be spelled out to determine their actual philosophical implications. In its most extensive definition, the materialist vision holds that a stone and a living organism do not differ in terms of their ultimate constitutive elements,

[1] For biologist Ernst Mayr (2004), the *physicalist's* conviction sustains that to obtain an exhaustive explanation of a phenomenon it is necessary to reduce it to its smallest components. You can obtain the explanation of a phenomenon only by analyzing its lowest organizational level. For Mayr *this* reduction is not only useless but also impossible. In my framework (6.3.2) there are validity conditions for this kind of reduction. For example, stability of the context or system is necessary for causal specificity of a part (6.4). In another publication, I say that the physicalist's reduction is not 'impossible' just 'impassable' (Bertolaso 2013a, ch. 4) to avoid the confusion with the analytical process. Analysis represents an important methodology in the study of complex systems. It always will. Reduction, in Mayr's view, is instead based on invalid hypotheses and should be eliminated from the vocabulary of science.

only in terms of *organizational properties* that determine the way in which these elements combine so as to give rise to structures of greater and greater complexity. This position might be compatible with both reductionist and anti-reductionist positions.[2]

Epistemological Reductionism and Its Evolution

Focusing first on epistemological (or inter-theoretical) reductionism in bio-medical sciences, it has been said (Ayala 1974; Rosenberg and McShea 2008) that there are two necessary and sufficient conditions for performing a reduction between a branch of science and another (Nagel 1961; Ayala 1968; Schaffner 1993): (1) *logical derivability*, according to which one has to show that all the experimental laws and theories of the former are a logical consequence of the theoretical constructs of the latter; and (2) *connectability*, meaning that all the technical terms used in the branch of science to be reduced to another are redefined with the terms used the latter. Thomas Nagel, in his work *The Structure of Science* (1961), proposed a form of inter-theoretic explanation characterized by a deductive derivation of the laws of the reduced theory from the reducing one, redefining the terms of the former on the basis of concepts that belong to the second. Famous exponents of reductionism in the 1970s and 1980s, such as Kenneth Schaffner (1976) and Michael Ruse (1976), noted that the hardest and most creative task of reductionism consisted precisely in establishing these connections, formulating principles that could *bridge* and tie together concepts belonging to both theories.

This approach led to extensive criticisms, beginning with David L. Hull, later followed by Philip Kitcher and others, and Schaffner himself.

Back in the 1990s, Schaffner published a paper on the "developmentalist challenge" (Schaffner 1993, 1998). The developmentalist challenge, often associated with the Developmental Systems Theory (DST, see Griffiths and Knight 1998), had stated in the 1980s that DNA does not contain a program for development and that genes are not the sole and main units of selection. Other factors, that are context-dependent, are relevant in the process of morphogenesis and evolution. The developmentalists' worry on one side was related with an experimental issue: behavioural geneticists can only study traits that are present and absent and thus miss variations.[3] On the other side, the question was whether "worm research can say anything useful about interesting research on human cognition" (Schaffner 2016, ch 4, 12).

---

[2] If we leave the question about materialism aside, we can focus on adherence to experimental physical data to define how the constitutive elements and the organizational properties of a living system might be understood (see Chap. 6). We can also concentrate on other aspects of the debate that regard the structure of scientific explanations: the identification of an explanatory level and the issues related with biological determinations (see Chap. 7).

[3] See Schaffner (2016), Chap. 4, for examples. I will not consider this argument in detail. However, I believe that clarifying the epistemological role of the context also contributes to clarifying other experimental issues related with the functional definition of explanatory parts. Presence, absence

The latter argument is based on experimental evidence showing how the reduction program in neurogenetics (Gilbert and Jorgensen 1998) fails because of emergent properties in neurons.

Schaffner (2006) recognized that when we try to capture what is going on in biological scientific practice, we can no longer apply the thesis that the reduction of one theory or one branch of science to another one, considered more fundamental (although inter-theoretical reduction still makes some sense in physics, Schaffner 2016). Schaffner introduced a different analysis of biological theory as a collection of *overlapping causal and inter-level models*.[4] Schaffner's aim was to understand the explanatory role of "simple" models in biological sciences that refer to the genotypes of animals. Acknowledging that biology lacks general laws, therefore it cannot apply a strict nomological-deductive model (ND-model) in the explanatory enterprise, Schaffner focused on what he called Causal Models (CMs) to account for the explanatory enterprise of biological sciences. The structure of biological knowledge, from both epistemic ad logic-of-explanation perspectives, is organized differently from what we find in standard accounts of the physical sciences (cf. Schaffner 2016, ch 4, 3). Although the assumption was still that a deductive process holds the explanatory power of an explanation, his conclusion was that the *explanantes* in physics are theories, while in biology they are models.

Following authorities in the scientific field, Schaffner first stated that model systems are a powerful heuristic for biology research,[5] but he never gave up the possibility to grasp, in epistemological terms, the nomological dimension of biological explanations,[6] going beyond mere heuristic considerations and assuming that models entail a level of idealization that justifies their use in different areas and fields of inquiry. The hope was that such models could eventually disclose conserved mechanisms that are applicable in general. Schaffner was cautious in saying that these kinds of continuities allow us to infer that simple model organisms can fully explain the behaviours of much more complex living beings; but he acknowledged that some kind of relevant information is driven by these explanatory models. This appeared plausible by considering that some molecular pathways are well conserved in nature at higher levels and that even the most complex behaviours in nature can be heavily affected by molecular factors. However, Schaffner recognized, this did not necessarily imply the possibility of reducing the explanation of higher-level properties' to explanations in terms of lower level properties. All we can say is that there is *continuity* and that the challenge is to understand better *what this continuity is* and how it is captured by reductive relationships.

---

and variations of traits belong, in fact, to this area of discussion as well. See in the book, for example, the context-dependent definition of parts (5.4), functions (7.1), and stem cells (7.2).

[4] This philosophical path would eventually lead to the Preferred Causal Model System approach described in Sects. 6.3 and 7.3 under the title "Conditions for Reduction".

[5] Cf. Schaffner (2016), Chapter 4.

[6] In Schaffner's account this nomological dimension relies upon the deductive character of explanations as assumed in the ND model.

Schaffner's support for these models in scientific practice immediately faced two challenges.[7]

The first challenge was posed by Gilbert and Jorgensen (1998). They rejected Schaffner's attempts to turn to the worm to look for a basis for behaviour in genes. They did so by claiming that deriving some kind of information from a simple model system like *C. elegans* (a worm widely used in biological studies for its simple structure of few and well differentiated cells) tracks down and eventually eliminates the richness of the developmentalist challenge (Gilbert and Jorgensen 1998, 259). For the developmentalist, "a gene may be an essential component of any behaviour, but it does not 'determine' it" (Gilbert and Jorgensen 1998, 260). In the context of a whole organism, single genes cannot determine discrete behaviours. "The predicted behavior of the animal does not emerge from the knowledge of genes and neuronal connections: the player acts independently of his or her teammates" (Gilbert and Jorgensen 1998, 261).[8]

The second challenge was the strong reaction by Developmental System Theory advocates against a reductionist account of their explanatory model. What the DST supporters claimed was that the relationship between genes and behaviour was not adequately captured by Schaffner's accounts. The *explanans* – genes in this case – cannot be considered a satisfactory explanation for an *explanandum* like a biological behavior. The *explanandum* exceeds its *explanans* even in the simplest models. In other words, this means that the normative role that the genes have in biological development does not follow the deductive structure of the ND model that still held in the explanatory character of Schaffner's accounts.[9] Genes as such have no normative priority over development. What DST theorists reacted against was Schaffner's (apparent) attempt *to resolve the everlasting tension between nature and nurture through a philosophy of scientific explanation focused only on their formal structure*. In fact, DST advocates reacted less against the formal structure of Schaffner's reductions, than against *the conceptual implications* that were drawn from reductions for the *relata*. In Sect. 6.2 and other parts of this book, I isolate two different epistemic dimensions: explanation and definition. A reductive explanation doesn't necessarily imply a reductive definition of the system or of the *relata*. On the contrary, *any reductive explanation entails, in my view, a non-reductive dimension which has to do with the definitions that are necessarily relative (epistemologically) and relational (ontologically).*

The (apparent) misunderstanding was immediately clarified by Schaffner who reframed the picture by distinguishing a heuristic and an epistemological issue:

---

[7] The two challenges are precisely related to *the kind of* continuity and reductive relation I am interested in understanding better, see Chaps. 5 and 6.

[8] The ultimate motivation for this emphasis is the fight against a deterministic philosophy, which might be dangerous for understanding organisms with consciousness, agency and the relevance of different environments (Gilbert and Jorgensen 1998, 263).

[9] Multiple realizability and variation in biological organisms are often used as an argument supporting the DST position described here. In my framework, however, the issue is deeper: it concerns how we know things as much as it concerns how biological things are organized (see Chaps. 5 and 6).

"Worm studies will not tell us anything about consciousness or intention or agency, for complexities exist in humans not found in simpler organisms, but some fundamental mechanisms can be found", and again "A worm is not a human, but worm studies, as well as other animals' (study), may offer important lessons about human psychopathology if yoked to other model systems" (Schaffner 2016, ch 4, p. 13). In this way, the opponents agreed upon the fact that, *to some extent*, genes control behaviors. What is at stake is "to what extent" and "what determines that 'to what extent'". Put in a different way, we should understand the "fundamentality" of models and the kind of information we get from simple models. Then the problem becomes to clarify the kind of information that these models bear and what universal features of biological Causal Model (CM) explanations convey.[10]

The Context Argument as a Possible Obstacle to Reductionism

It is worthwhile to go deeper into the context argument of these debates. Anti-reductionist DST positions claim that the context has a relevant role in explanatory accounts, while reductionist supporters tend to deny it or to confine the argument to a pragmatic level, understood as the actual interest of scientists at a given time.

To some extent we can say that geneticists and developmentalists stress different aspects of the same picture. The geneticists' emphasis is on the genetic parts. However, how genotype and phenotype are related is still an unresolved issue. The developmentalists' emphasis is on the context[11]: "genes have little meaning … per se, only in context with other genes and in the environment that is cellular, extracellular and extraorganismic" (Schaffner 1998, quoted in Griffiths and Knight 1998). That not everything matters is commonly accepted. But why some traits matter more than others and why the context matters at all is not yet clear. As Schaffner puts it:

> The dangers of DST in its present form, as I see it, is that it gives too much to "context" […] and needs to formulate its categories of interactions more clearly […]. It is not helpful to assert that everything interacts with everything else, but that could be a problem for DST unless it provides us with some form of prioritized ontology (Schaffner 2016, ch 4, 15).

"Prioritized ontology" is also what developmentalists refer to,[12] but we are not able to explain in which sense – for example – genes and environment can have a

[10] The specificity of this scientific information also refers to the relationship between the definition of the parts and the actual instantiation in biological explanations. This can be easily read in the relationship between genotype and phenotype as well. The point is to understand how genes are defined in order to explain a higher-level behavior of biological parts, i.e. to spell out the traditional problem of the relationship between the genotype and the phenotype (see Sect. 6.4, "Stability Wins over Specificity").

[11] Indeed, the developmentalist challenge has also been addressed as 'contextualism'.

[12] What is missing here, to me, is that a "priorized ontology" is at hand only when considering how things work from the perspective of the context dependency of biological explanations (see Sect. 5.2.5).

priority. The same tension is true, in explanatory terms, in the parts and wholes discussion (see Sect. 5.2).

The way out for Schaffner and others is to recognize that genes, causally, have *parity* with other molecules as several necessary and jointly sufficient conditions (to produce traits). However, epistemically and heuristically, genes do have a *primus intra pares status* (Schaffner 2016, Ch 2, 51).[13] Nevertheless, this issue puts Schaffner in a peculiar position: as the previous debate shows, how genetic and environmental components are related is not straightforwardly obvious when the claim is that their explanatory power is due to the deductive relationship that involves them.

This point is easier to understand, also in the light of the examples from cancer biology, if we consider that what is really at stake in "prioritized ontology" is not a traditional ontological priority (what really exists) but an *onto-epistemological priority* (how we conceptualize what is relevant in explanatory terms through causal relations). There is a non-symmetric dependence in the reductive relationship that defines the *explanans* and the *explananda* of a reduction, which the logical framework we inherited from the ND model is not able to capture. For example, we can consider that there are genes, but their definitions depend on the identification and previous characterization of the biological behaviour to be explained. As we discuss all over the book, in the part-whole language, the part and the whole are not 'parts' in the same sense (Sect. 5.2).

In book Chap. 1 we see the problems posed by the timing of the neoplastic process and the multiplicity of the internal and external causal factors eventually ascribable to un-coupled dynamics and tumor heterogeneity. In Chap. 2 we encountered the Somatic Mutation Theory and more complex Cell-Centred models, and we also met the promises, doubts and delusions of advocates such as Weinberg, the same scientist who wrote the aphorism "Anything found to be true of *E. Coli* must also be true of elephants" (Weinberg 2006), an eloquent example of strong multidimensional reductionism.

According to the Standford Encyclopedia of Philosophy (2012) there are three main problems for reductionism in biology: *context dependency, multiple realizability* and *temporality, intrinsicality and representation* of biological systems. I am not able to address all these issues systematically and to highlight criticisms of reductionism, but I think I have subsumed all of them by focusing on *contextuality*

[13] This claim supported by the author, was also referred to as indivisibility, i.e. the idea that individual genetic and environmental causes cannot be identified by separable effects on the phenotype, and the effect of all causal factors are, in some way, context dependent. Schaffner will never spell out the insight he got through this notion of indivisibility. He instead prefers to adopt a weak notion of emergence where the kind of unpredictability that "means that from total information about genes and environment, we cannot predict an organism's traits: they are, accordingly, emergent" (Schaffner 2016, ch 2, 49). Coherently, he did say that genetic determinism should be defended, but that developmental noise (an argument for strong emergence in developmental theories) should be viewed with suspicion. In his framework the only way to understand heterogeneity in the behavior of biological "parts" of a "whole" was in terms of noise. This is so because the premises of a deductive model logically entail the terms of deduction.

all over the book, with empirical examples in Chaps. 1, 2, 3 and 4, and especially with theoretical discussion in Chaps. 5 and 6. The context, in fact, is relevant in methodological, epistemological and ontological terms. Typical examples of context-dependency are the multiplicity of functions of proteins, the redundancy of the genetic code, according to which many different sequences can codify for the same gene, etc. But, although the context argument is mainly related to the one-to-many issue, other aspects such as temporality (and its sequential and hierarchical dimensions) and intrinsicality (i.e. how what is "internal" and what is "external" are explicitly or implicitly distinguished in reductive explanations) of biological systems recall the context argument in different ways.

Schaffner's "simple system" strategy appears as a variant of the mechanistic and systemic approaches but with an important difference. Context sensitivity, which is a problem for strongly mechanistic committed accounts, is recovered in the Causal Model (CM) through the concept of Connectability Assumptions (CAs), i.e. conceptual assumptions that, once specified, permit to create a bridge between macrodescriptions and microdescriptions (e.g. a behaviour and genes' activity). All these insights are taken up in the ideas about *mesoscopic reasoning* (Sect. 6.3), reduction as identification of mesoscopic levels (Sect. 6.3.1), conditions for reduction (Sect. 6.3.2) and related issues. Within this framework, I suggest that the role and status of Connectability Assumptions (CAs) should be reconsidered. Schaffner rightly assumes that they can be *either* causal sequences *or* identities. However, the specificity of the system calls for both features, and the continuity of levels and of degrees of complexity in the natural world brings them into the picture, so that reductions are not only "possible", they are also required in science. CAs don't seem to constitute a condition for reductions, but to integrate them as a token of the non-reductive dimension of any biological explanation.

Mechanism

Let us now briefly address the question about the relevance of *mechanistic explanation* in biological sciences as in the last decade a substantial part of philosophical literature has been centred on this point.

In its origins, mechanism attempts to explain the physical world by the movement of inert bodies that are pushed or pulled through direct or indirect physical contact with other bodies. Its proponents usually hold that local motion is the only real motion, and that a body is maintained in such motion by its own inertia or impetus. Mechanism as a metaphysical doctrine is often associated with the view that everything can be understood in purely by quantitative and geometrical principles (extension and motion), thereby giving mathematics primacy in physical science.

The adoption of a mechanistic approach in biological experimental practice can be related with its commitment to avoid references to factors that either (1) fail to increase the simplicity and scope or fecundity of biological theory; (2) fail to increase the precision and ease with which biological theory may be empirically confirmed; or (3) are conceivable only by "anthropomorphic empathy", in terms of such concepts as will, desire, and urge.

The Twentieth century has been characterized by the risk of transforming those methodological recommendations into a philosophical and ontological view of the biological world and of the human being, according to which living systems can be conceived in mechanical terms, i.e. as assemblies of interacting parts arranged in such a way that their combined operation results in predetermined outcomes (Bertolaso 2013c). However, the mechanistic approach is fundamentally compatible with the recognition of system laws that elude the reduction to the laws that govern physical and chemical processes. Mechanism may also admit the unpredictable course of evolutionary history and the irreducibility of the principles of natural selection and biological behaviours. Biology and physics are not required to be identical or even similar in their concepts or in their basic laws.[14]

A mechanistic explanation identifies parts and their organization showing how the behaviour of the machine is a consequence of the parts and their organization. In this book I adopt this classical perspective when talking about mechanism (e.g. Sects. 2.9 and 2.10). I relate mechanism with the *mereological and causal* features of mechanistic accounts that have been elaborated more recently (Sects. 2.9 and 4.3.3). Mechanisms have been defined either in mereological or explanatory terms: they consist of "entities and activities organized in the production of regular changes from start or set up conditions to finish or termination conditions" (Machamer et al. 2000, p. 3) or "an explanation of systemic behavior in terms of the behaviors of the constituent parts within the systemic context" (Richardson and Stephan 2007, p. 139). A more comprehensive definition of mechanisms has been offered by Bechtel:

> What unites one set of parts and operations into a given mechanism is their organization and their orchestrated functioning in producing a particular phenomenon. [...] A mechanism is typically not just a collection of independent parts, each carrying out its operation in isolation. Rather, parts and operations are generally integrated into a cohesive, functioning system (Bechtel 2006, p. 29–33)

Once again, mechanisms lead to the issue of how systems are defined. I dealt with this point in Chaps. 5 and 6.

Different accounts of mechanisms have emerged when considering that "[t]he complexity of biological and biomedical phenomena is also seen as particularly challenging, especially given the rise of systemic/integrative approaches wishing to understand how entities and processes at different levels of organization, ranging from genes and cells to organisms, populations and ecosystems, shape and construct each other" (Leonelli 2012, Introduction). Extensive literature on this topic is available since some authors, like W. Wimsatt, M. Ruse, and D. Hull, highlighted the limits of Nagels' account of scientific explanation in understanding what was really going on in biological science. However, theories of reduction and explanatory

[14] So understood, a mechanistic approach is not necessarily incompatible with philosophies of nature that recognize the independent validity of non philosophical investigation of biological processes or that speculate about the human or metaphysical significance of such scientific inquiry.

reduction have gone hand in hand since then (Brigandt 2006). The proper concept of mechanisms is considered as an alternative to law-based approaches to explanation and reduction.[15] With the *New Mechanistic Program*, even more attention was paid to mechanistic approaches in biological sciences. The influential paper written by Machamer, Darden, and Craver in 2002 offers a general characterization of 'mechanism' that attempts to capture the way scientists use this word and to show the ways in which mechanisms are involved in the explanation of phenomena. Examples taken from different domains in life sciences, from molecular biology to physiology and neurobiology, have been used to support mechanistic accounts. The argument was that scientists often make use of a mechanistic vocabulary, therefore a philosophical reflection on the term 'mechanism' is worthwhile. And in fact it is. However, I tried to show in this book that if we clarify what mechanistic talk *really* look like in science, we grasp the implications of its intrinsic structure, and we are led to a Dynamic and Relational View which is very different from the usual mechanistic accounts.

According to Machamer, Darden and Craver's (henceforth, MDC) fortunate definition of mechanism, mechanisms are "entities and activities organized such that they are productive of regular changes from start or set-up to finish or termination conditions" (Machamer et al. 2000, p. 3). To bring the dynamic feature of biological systems into the picture, the traditional MDC model has been revised. Richardson and Stephan define a mechanistic explanation as something in the sense advanced by MDC but enriching it with Kauffman's terms (1970). Mechanisms can be seen as "an articulation of parts explanation", so that a mechanism is "an explanation of systemic behavior in terms of the behaviors of the constituent parts within the systemic context" (Richardson and Stephan 2007, p. 139). The reference to Kauffman's work adds to the picture a clear reference to the regulatory dimension that a mechanistic explanation is meant to capture. For Kauffman, in fact, a decomposition of the system into parts is conditional on what is seen as the "goal" of the system, what the system is doing. The same attempt, as reviewed by these authors (Richardson and Stephan 2007; Bechtel and Richardson 2010), is entailed in Glennan's account of mechanism for behavior as "a complex system that produces that behavior by the interaction of a number of parts, where the interactions between parts can be characterized by direct, invariant, change-relating generalizations" (2000, S344). What features of the behaviour of the parts and of the whole are explanatorily relevant in the former definition, and what is the meaning of 'generalizations' in the latter one, remain to be clarified.

Discussion about mechanisms is often centred on what mechanisms are and how mechanisms explain. *What* they explain is often overlooked. Avoiding a discussion on this point might imply that mechanisms that *explain* something also *define* it. In fact, routine accounts consider a system as just an aggregation of its fundamental parts, so that parts are causally relevant by virtue of their intrinsic properties (see also Bechtel and Richardson 2010). Neither a further definition of their features, nor a further discussion about their epistemological role in the structure of a scientific

[15] This point is made by Schaffner (1993, Ch. 3 and 2006, p. 398).

explanation, is required. *Ad hoc* arguments about the context relevance are sometimes included. Nevertheless, in scientific practice mechanisms are used to account for *emergent* properties. I take emergent property in its broader meaning, i.e. a higher property that is usually perceived as a behaviour or a stable functional – or end – state of the system of interest. From a scientific point of view, we need to explain robust phenomena, i.e. the organizational dynamic of an emergent pattern. This is *what actually resists* this kind of reductionism in biological sciences. From a philosophical point of view, we need to understand how different levels of biological organization are actually inter-regulated. This implies understanding where the nomological dimension – that any regulatory process entails – has to be searched. The regulatory and inter-level features of biological dynamics become the real challenge of any explanatory account of biological processes and phenomena. They show, in fact, a synchronic and reflexive dimension that might be contemporarily captured by the explanatory account.

The Ambiguity of Mechanism in Face of Organismic Regulation

Craver's characterization of mechanistic explanations – as multilevel causal explanations that "explain by showing *how* an event fits into a *causal* nexus" (2001, p. 68) – leaves open the interesting idea that multilevel explanations are constitutive and need not refer exclusively to proximate causes (Richardson and Stephan 2007). If the mechanistic dimension of the explanatory account takes over the causal explanation in traditional efficient terms, the question is: what is determining what the relevant entities and activities in a mechanism are?

Appealing to inter-level experiments as "tools for determining what the relevant entities and activities in a mechanism are, for determining how they are nested in a multilevel hierarchy, and for showing how a given component is integrated within its mechanistic context" (Darden and Craver 2009, p. 3) does not explain the *relative dependence* of the parts on the wholes, and does not overcome the ambiguity that the definition of mechanistic explanations entails. The answer to "Why does X behave this way?" in mechanistic terms, i.e. "because there are mechanisms", is always possible, but hardly useful for understanding what is going on in biological regulatory processes, without a clarification of how we get to those mechanisms, i.e. what is *possible*. Multilevel experiments are performed in biological sciences, but their explanatory power cannot rely just on the identification of further mechanisms.

What follows is that the commonly accepted idea that "a mechanism must trace how a phenomenon is caused using the objects and activities appropriate to the field and must account for each step in this process, *leaving no gaps unaccounted for*" (Delehanty 2005, p. 719, my emphasis) is still challenged by the characterization of "the objects and activities appropriate to the field". These are, in fact, not autonomous from an epistemological point of view, and a gap has to be filled anyway. This has been called elsewhere the problem of the "system level understanding" (Kitano 2002) that characterizes scientific explanations in biological sciences, strictly related to the epistemological status of generalizations in biological explanations. As Richardson and Stephan put it: "Whether or not this is the only route to explain complex systems, genuine understanding is reached when we are able to describe a process, or a complex system, with a grade of resolution that allows us to see the relevant components 'at work'" (Richardson and Stephan 2007, p. 142).

These considerations cast serious doubts on a strong and simple mechanistic reductionist account of biological phenomena. From a phenomenological point of view in mechanism no entities with a causal priority are defined. There is causal equivalence of the component of the system. In fact, what emerges from a deeper consideration of mechanistic accounts in the scientific literature, is a definition of mechanisms where almost all references to mereological and static features disappear (see Bechtel's quote above).

If relations among parts *weren't* relevant in nomological terms,[16] so that parts could be considered homogeneous and treated in terms of average, reductionism in the traditional sense would work and this has been, in my view, one of the major insights that we have ever had in science (cf. gas laws). Problems arise when we assume those requirements (i.e. intrinsic equivalence of constituent parts of a system and completeness of explanation in terms of parts) as intrinsic conditions of any scientific explanation or theory. The nature of biological theories has been misconceived. Reductionism in those terms is simply "impassible" (Bertolaso 2013a) because it does not equip us with the tools to grasp one of the intrinsic features of biological complexity to which it is tied.[17] The same happens with strong mechanistic claims: we agree that the context is relevant but there are no tools to explain in what sense this is the case. It often seems to be an *ad hoc* assumption or a statement that arises only from pragmatic concerns or aims. This is the case when supporters of mechanistic explanations advocate for an integrative approach to inter-level phenomena (Darden and Craver 2009).[18]

---

[16] What should be considered nomological would require a deeper reflection as, from what we are saying, its meaning is changing in this analysis. At the moment I leave this point open. An exhaustive discussion of it is beyond the aim of this volume.

[17] One could say that reduction without specifying the contextual interactions is impassible, but when inter-relations are provided, an explanation or more sophisticated reduction (like in an innocuous emergence) works. The answer is that this would not be the case if, as I am suggesting, the PCMS element, to which the specification of the contextual interactions belongs, implies a conceptual change in the definition of the *relata* (from genes to cells to, eventually, the tissue organization in the biology of cancer). A series of PCMS can still be identified but their *explananda* will be different. Further discussion of this point might require a clarification of the historical and evolutionary dimension of the biological systems that is beyond our scope.

[18] We have to highlight here that Craver has moved towards a position which actually converge with the account of mechanistic explanations I would defend (Craver 2009). In particular I would highlight for the readers the common use we make of the notion of 'relevance' in explanatory terms (*ibidem*, p. 590).

Mechanistic reductionism, then, does not appeal to a 'nothing but' or 'lowest level' claim in its explanatory accounts, but denies in practice that something really new happens in nature as all can be eventually explained in mechanistic terms: the 'nothing new' issue. The difference between explaining *how* something does what it does and explaining *what* it does through a mere specification of the mechanisms at work is still an unresolved challenge for the reductionist explanation.

It does not seem that the integration of the traditional MDC account through the emphasis on parts' and wholes' behaviour and context is neutral with respect to the explanatory role of mechanisms themselves. This is equivalent to asking what pertains to the system that confers to the identified mechanism an explanatory power, or how the dynamic feature of a biological system influences the structure of the mechanistic accounts.

Schaffner's Causal Model (CM) account addresses similarly this same problem, i.e. what generalizations the inter-level regulation of biological systems implies. The discussion about the partiality of CM models and of their causal account might be finally related to the generalizations required to account for biological processes and their pathological features. As we will see now, such enterprise is related with the 'novelty' of higher-level properties. From a methodological point of view, that 'something new' first requires to identify a level of analysis where parts, their interactions and their contextual dependence can be integrated into an explanatory model, so that some higher-level features can be eventually described in terms of mechanisms. It is the identification of the explicitly relevant object that constitutes the real challenge of reduction for biology.

Dupré and Keller on Reductionism and Anti-reductionism

Three issues are usually at stake in the reductionist-anti-reductionist debate: interactions vs. relations; top-down vs. bottom-up causality; identity of the part and the whole. The debate between philosophers John Dupré (2010) and Evelyn Fox Keller (2010) is representative of the debate between anti-reductionsm (Dupré) and reductionism (Keller) in the philosophical literature on these issues, and relevant to some arguments I have been presenting.

Keller and Dupré agree that "[w]hatever the meaning of *fundamental* in biology, it clearly *cannot be equated with simple*, nor is it at all obvious that it is common to all biological entities" (Fox Keller 2010, p. 20, my emphasis).

They diverge when Keller claims that biological explanations "require *nothing more than the working of physics and chemistry*" and that "analysis of the specific physical and chemical phenomenology involved in biological processes should, in principle, suffice for an understanding of what endows biological systems with the properties of life" (Keller 2010, p. 21, my emphasis).

Dupré challenges the reductionist principle (RP), i.e. the conviction that "if we knew everything about the chemicals that make up a lynx, and the way they are assembled into cells, organs, and so on, we would, in principle, know everything about the lynx" (Dupré 2010, p. 34). Anti-reductionists do not agree that systems are "nothing but a collection of physical parts *assembled in a certain way"* (Dupré 2010, p. 34) assuming that this "in a certain way" is neutral (in explanatory terms) with regard to the differences between living and non-living systems.

A non-symmetric relationship results between the reductionist and anti-reductionist position. Dupré, discussing the statement that "systems are 'nothing but' a collection of physical parts assembled in a certain way", distinguishes two different questions: one about the *definition* of the *relata*, and the other about *in what way* a reduction can be realized. Keller, in her discourse, focuses on 'in what way' a reduction of biological explanations can be performed. Without contradiction with what they have been claiming, Dupré and Keller can really agree upon the fact that there is a "*dependence of the identity of parts*, and the interactions among them, on higher-order effects" (Dupré, p. 37, my emphasis). In both cases (reductionist and anti-reductionist) there is something that exceeds the assumed stance and relates it to its opposite.

The central issue is related to how parts are organized in a whole. Anti-reductionists would say that new relations arise: although "the emergent features of a whole or complex system are not completely independent of those of its parts since they 'emerge from' those parts, the notion of emergence nonetheless implies that, in some significant way, they *go beyond* the features of those parts" (Machamer and Silberstein 2002, p. 1404). For anti-reductionists, "assembled in a certain way" adds something to the definition of the parts and of their interactions. For reductionists the claim that "the emergent features of a whole or complex are not completely independent of those of its parts" is, on the other hand, the only perspective that is worthwhile to be taken into account in the scientific inquiry. The 'dependency' anti-reductionists refer to is qualitative while in reductionist accounts it is mainly quantitative (or at least they don't have any epistemological tool to explain in which sense it might be qualitatively different).

Although Dupré initially poses the discussion in epistemological terms – so that "properties of constituents cannot themselves be fully understood without a characterization of the larger system of which they are part" (Dupré 2010, p. 32) – the final issue is related with the concept of biological properties. In particular, Dupré articulates the notion of properties as follows: (a) properties of interest in complex systems are *capacities*; (b) capacities are defined in terms of *behaviour* that is their characteristic exercise. "Moreover, capacities, I argue, are jointly determined by intrinsic features of an entity and by features of its environment" (Dupré 2010, p. 45). Capacities are always context-dependent for Dupré: the *context* refers "to features of an object's environment that are necessary to confer on the object a particular capacity (…). *Interactions* are simply the *exercise of such capacities* with relation to some other entity that will presumably constitute all or part of that context" (Dupré 2010, 45, my emphasis).

Instead, Keller thinks that the distinction between *context* and *interactions* is artificial in Dupré's discourse: "context is simply all those other factors/molecules whose interactions with the object or system in question have not been made explicit and, hence, have not been included in the description" (Keller 2010, 30). Keller stresses *interactions* among parts, leaving apart the definition of the system. In this way she focuses on the working of the defined system and she can claim: "[as a materialist] I am committed to the position that all biological phenomena, including evolution, require nothing more than the working of physics and chemistry" (2010, 21). Under this position, the *notion of function* is minimalist, a simple feedback mechanism (Keller 2010, 24; cfr. 6.6): it presupposes the existence of single celled organisms, of stable and autonomous cells capable of dividing. It is accepted that functions and functional accounts endow biological systems, but a deeper discussion is usually avoided, while functional arguments are included in the explanatory accounts.

Dupré opens the discussion to the possibility of distinguishing two different dimensions of biological explanations: "I would say that the project of characterizing the entity, which I have said requires reference to the context, and the project of describing what, on a particular occasion, it does, namely interacts, are distinct activities" (Dupré 2010, p. 45). The capacities of an object of inquiry are thus not merely consequences of its molecular constitution, but are simultaneously determined by the systems of which the object are parts (Dupré 2010, p. 39).[19]

The discussion about properties (above defined in terms of capacities) is integrated by the issues of *emergence and downward causation*. In Dupré's account, in fact, (dispositional) *properties are relational* (i.e., they cannot be reduced to any information about the parts) and the relevance of the context is double: on one hand, it is related with the *identification* of the explanatory parts, on the other hand it plays some crucial role in the *maintenance of the identity* of the parts itself. Keller says that she uses top-down causation "to refer to a wide range of influences that global properties have on the parts, including not only their activity (in the case of genes), but their very identity. (…) Indeed, the very definition of what (if anything) a gene *is* depends on the properties of the cell in which the DNA is embedded" (Keller 2010, p. 30, my emphasis).

Keller claims that "the biological explanations require nothing more than the working of physics and chemistry" and that "analysis of the specific physical and chemical phenomenology involved in biological processes should, in principle, suffice for an understanding of what endows biological systems with the properties of life" (Keller 2010, p. 21). The epistemological status of concepts here expressed as "working" and "phenomenology" remains open to a broader discussion. I do believe that this discussion coincides with questioning how *relata* are identified.

From a methodological point of view this is embodied in the question that Keller poses as follows: "What kinds of analysis, in fact, are required?" (*ibidem*). Her answer is that Systems Biology, considered as an analytic technique, is useful to move beyond a list of parts. For Keller this implies that "we are beginning to acquire

[19] A claim present in my Framework, stronger than Dupre's one about capacities, is that *the conceptualization* of a part, as an object of inquiry and analysed in terms of constituent parts, *requires* to define the *context* with which it interacts. This stronger claim is very important if we apply it, for example, the ways that we divide the genome into functional parts (genes!), or to the definition of cancer cells.

the kinds of understanding that would permit the formulation of biological explanations in terms of physical and chemical processes" (Keller 2010, p. 22). In my view, the picture is more complex than that: *the reasons that lead* to the establishment of the new methodology that we call Systems Biology have to be considered (cf. Chap. 3), and these reasons lie in the dependency of the identity of parts and of their interactions on higher levels effects: "Systems biology of any persuasion has to demonstrate that when single components come together and form a system, they engage in novel behaviour and produce novel phenomena by the system itself constraining the components" (O'Malley and Dupré 2005, 1273, my emphasis). Explaining the kind of dependencies of parts from the whole would eventually clarify the issue of higher-level properties; clarifying the epistemological issues posed by the causal relevance of the higherlevel properties, will give us insights about the ontological structure of a biological organization.

Reductionist supporters seem to believe that the whole game relies on proving the possibility of theoretical reduction; so that they mainly concentrate on 'in what way' the reductionist relation is built. On the other hand anti-reductionists are more sensitive to the dynamic properties of systems and their implication in causal and explanatory terms. As Dupré puts it:

> I want to deny that the behavior of the whole is fully determined by the behavior of, and interactions between, the parts. And hence, the elements of behavior that are not so determined are what we don't know when we know everything about the parts and the way they are assembled [so that] the fact that biology – a science – works with concepts that depend on the larger systems of which they are part, as well as on their constituents, is a fatal objection to the claim that 'it is possible to reduce biological explanations to explanations in chemistry and/or physics' (Dupré 2010, p. 38).

Emphasis is on the conceptual definition of systems and on the epistemological perspective to adopt in order to account for such dynamic properties.

What emerges is not a contrast between reductionist and anti-reductionist positions but a question about the philosophical foundations of an anti-reductionist position and the consistency of a non-reductionist approach in life sciences. In what sense might reductionist and anti-reductionist positions imply each other? Why do they seem to reach out for each other? What is the real issue that divides them and eventually grounds their philosophical incompatibility? What is *the relationship* between reductionist and anti-reductionist positions?

Easy simplifications are not a good way of proceeding and the current problem is not just a matter of deciding what reductionism is (cf. Rosenberg 2007). What is eventually at stake in the relationship between reductionism and anti-reductionism in biological sciences is the view on *how biological systems are structured and what implications this has for science*. Understanding how reductions are performed and how they work in science is but a first step (Sect. 6.3).

When looking for the philosophical foundations of non-reductionist positions, ontological and epistemological questions converge only apparently on the epistemological one, when the debate on reductionism is centered on the relationship among the parts and the whole to which they belong. They seem to structure the debate between reductionism and anti-reductionism in terms of the "in principle"[20] issues like in the above quote of Dupré's. This however, does not come without a price. The above-formulated question on the relation between ontological, epistemological and methodological reductionism remains open. The convergence of these three dimensions and its justification can subsist if we consider that the scientific method proceeds through the decomposition of the systems in parts, the interactions of which can determine a particular phenotype. Therefore epistemological considerations have an obvious priority in the analysis of scientific practice, but ontological questions remain open beyond the definition of the explanatory relevant biological interactions.

In these terms, the challenge formulated by Dupré meets the terms of the discussion about reductionism posed by Silberstein (Machamer and Silberstein 2002), who distinguishes issues related with the "relata" of a reductionist or an anti-reductionist account and "in what way" a reductive relationship is construed. He states that the best reason for believing in reductionism is an *acceptance of mereological and/or nomological supervenience* based, in large part, on successful inter-theoretic reduction or epistemological reduction that are the two issues challenged up to now.

Emergence and Mereology

The question about the possibility of reduction of biological explanations eventually leads to the hard core of the reductionist debate in biology: *biological emergent properties* and the discussion about what the *context* matters. I would say that a relational approach, as developed in this volume, helps to overcome these issues. Arguments about emergence are not pivotal any more in the debate. The search for what is relevant in explanatory terms, in fact, shifts the focus on the *structure of biological explanations* away from how reductions are performed. Reflection is therefore on the kind of causal dependencies that are at work in the structuring processes of the natural world, no more on the (causal) relevance of higher (emergent) properties in biological systems. This latter debate was still based on the mereological-mechanistic frame adopted by mechanistic accounts.

The concept of "supervenient" is rooted in the one of "emergence", and often used by philosophers in order to indicate the derivation of biological kinds on the

[20] These obstacles distinguish a different kind of anti-reductionism or emergentism, i.e. the belief that there are properties that could not have been predicted and explained (even in principle) from a complete knowledge of the constituents and their parts. Prediction and explanation are equally used in this kind of claims. However we will focus on explanation that is strongly related to philosophical issues, not only in the sense of the epistemology that looks at 'how' we know but to the conditions of validity of these claims.

basis of disjunctions of physical kinds. Supervenience is often use to conceptualize the natural selection of functions (understood as selected effects), which is blind with respect to structure, implying that all biological properties can be realized in a multiplicity of ways, and that there is a "principle of neutrality of the substratum" (Dennett 1996) for which "the power of the procedure is due to its logical structure".[21] However this philosophical position arises in my opinion from the analysis of *diachronic emergence* and from the assumption that this is the main issue at stake in order to understand emergent phenomena. The difference between the biological phenomena that are really emergent from those that derive from an erroneous application of the concept of "supervenient" (in the field, for example, of evolutionary biology) implies something more, as we see in Sect. 5.2.3 with the discussion of *reflexive and synchronic features* of biological systems. Therefore, physicalism and supervenience as ontological and epistemological theses about the biological world seem weak compared to the evidence that living systems are not closed but *open systems* (Soto et al. 2008b).[22] This implies, as working hypothesis at this level, that other explicative dimensions should be admitted, i.e. *other causal categories are necessary in order to grasp if, and in what terms, a discussion of biological dynamics based on supervenience is satisfactory*. Within a different perspective, Rosenberg highlights this point in an interesting way: "There is of course no trouble identifying 'tokens' – particular bits of matter we can point to – of genes with particular 'tokens'

[21] Both premises would be coherent with the conclusion that what makes of two genetic sequences the same genes is, for example, their functional role (Rosemberg 2008). Now emergence can be accepted from this point of view as an epistemic property that implies unpredictability starting from the lower levels as long as it is due to a "mereological supervenience": this seems to be the case also for Rosenberg and Kaplan (Rosenberg and Kaplan 2005). We understand, for "mereological supervenience" "…systems with an identical total microstructural property have all over properties in common. Equivalently, all properties of a physical system supervene on, or are determined by, its microstructural property" (Kim 1999). In this perspective, emergence cannot be an ontological property, a property that implies any sort of quantitative novelty. This perspective is finally compatible with the idea that there is a "complete micro structural description" (Kim 1999) of the system. The macroscopic properties can be considered unpredictable and in a certain sense unexplainable in so far as the relation between a property *M* and a property *P* is not univocal. For what regards supervenience and divergence, *M* depends on *P* o a derivative of *P*, but this only implies that, given a certain physical system, one should always have the same "emergent" macroscopic properties, and therefore physical facts would fix all the facts. "A thesis of the supervenience of the biological on the physical asserts that, however inaccessible are principles connecting lower levels to higher levels, nevertheless, the biological depends on the physical in the sense that for any biological system there is a physical state that constitutes it, and wherever we were to find an identical physical state we would find an identical biological state" (Dupré 2010, p. 45).

[22] I will not follow this line of analysis. It dates back to cybernetics, and other related systems theories, but I believe that the relational epistemological framework I am suggesting in this volume could be useful for a revision of the problems that this account of openness of living systems also originates.

of their molecular constituents. But token identities will not suffice for reduction, even if they are enough for physicalism to be true" (Rosenberg 2007, p. 122).

Another intermediate issue discussed extensively in philosophy of biology is the possibility to reduce discussion about emergent properties and reductionism to the question on *multiple realizability* (or many-one relationship), i.e. the fact that practically all biological properties can be realized in a multiplicity of ways. Authors like David Hull widely dealt with it (Hull 1976; Sober 2010). The complexity of functions revealed by the discovery of sequences and regulatory sites of genes, of introns and exons, of the translational and post-transcriptional modifications, of promoters, operons, open reading frame, junk DNA, transposons, virus DNA, etc. contributed to reinforcing the reflection on multiple realizability, which often appears as an argument against reductionism as it is difficult to define genes connected to specific cellular functions in terms of their DNA molecular structure. In my opinion, however, Sober is right when he says that the multi-realizability argument is not a definitive argument against reductionism (Sober 2010). It is instead an argument for discussing the relative autonomy of levels of biological organization or, as I call them, *functional emergences* or (even better) *functionalities.*[23]

This meets Dupré's anti-reductionist concern about biological determinism:

> I want to deny that the behavior of the whole is fully determined by the behavior of, and interactions between, the parts. And hence, the elements of behavior that are not so determined are what we don't know when we know everything about the parts and the way they are assembled (Dupré 2010, 35).

Here, in my opinion, is the relevant point. It does seem questionable that a mechanistic explanatory reduction, although partial, of biological behaviour is reasonably sufficient to account for the explanatory feature of biological models, without any remainder. In any case, acknowledging emergent properties in mereological terms does not resolve the question (see Sect. 5.2.3).

[23] From this perspective I understand better in which sense the discussion on the existence of laws in biology and specificity of these laws followed as a necessity (Rosenberg 2007; Ayala and Arp 2010; Mitchell 2009, etc.). From the point of view of cancer biology, this seems to be a speculative rather than practical problem, and therefore of minor interest for both philosophy and science. The irreducibility of mendelian genetics to molecular genetics, on the basis of the above mentioned data, does not allow to attribute the properties typical of a law to the "Mendelian laws", but to consider them as generalizations, descriptions of a large number of particular facts, valid only in a specific context. Historically, the question of autonomy of biological sciences with respect to other sciences was a derivation of this same problem. "[T]here are no laws of biology to be reduced to laws of molecular biology, and indeed that there are no laws of molecular biology, can be shown by the same considerations that explain why genes and DNA cannot satisfy reduction's criterion of connection" (Rosenberg 2007, p. 122). More in the Appendix.

Bibliography

Abbs, S., Bussoli, T., & Kavalier, F. (2004). Nature encyclopaedia of the human genome. *BJM, 328*, 172.

Agazzi, E. (1978). Systems theory and the problem of reductionism. *Erkenntnis, 12*, 339–350.

Ailles, L. E., & Weissman, I. L. (2007). Cancer stem cells in solid tumors. *Current Opinion in Biotechnology, 18*, 460–466.

Alberts, B., Johnson, A., Lewis, J., Raff, M., Roberts, K., & Walter, P. (2002). *Molecular biology of the cell*. New York: Garland Science.

Ames, B., Durston, W. E., Yamaski, E., & Lee, F. D. (1973). Carcinogens are mutagens: A simple test system combining liver homogenates for activation and bacteria for detection. *Proceedings of the National Academy of Sciences of the United States of America, 70*, 2281–2285.

Ao, P. (2007). Orders of magnitude change in phenotype rate caused by mutation. *Cell Oncology, 29*, 67–69; author reply on *Cell Oncol, 29*(1), 71–72 Comment on: Duesberg, P., Li, R., Fabarius, A., & Hehlmann, R. (2005). The chromosomal basis of cancer. *Cell Oncology, 27*, 293–318.

Ao, P., Galas, D., Hood, L., & Zhu, X. (2008). Cancer as robust intrinsic state of endogenous molecular-cellular network shaped by evolution. *Medical Hypotheses, 70*, 678–684. Epub 2007 Sep 4.

Aranda-Anzaldo, A. (2002). Understanding cancer as a formless phenomenon. *Medical Hypotheses, 59*, 68–75.

Artiga, M. (2011). Re-organizing organizational accounts of function. *Applied Ontology, 6*, 105–124.

Artigas, M. (1992). *La intelgibilidad de la naturalezza*. Pamplona: EUNSA.

Ashby, J., & Purchase, I. F. (1988). Reflections on the declining ability of the Salmonella assay to detect rodent carcinogens as positive. *Mutation Research, 205*, 51–58.

Ashkenazi, A., & Dixit, V. M. (1998). Death receptors: Signaling and modulation. *Science, 281*, 1305–1316.

Ayala, F. J. (1968). Biology as an autonomous science. *American Scientist, 56*, 207–221.

Ayala, F. J. (1974). Introduction. In F. J. Ayala & T. Dobzhansky (Eds.), *Studies in the philosophy of biology* (pp. vii–xvi). Berkeley/Los Angeles: University of California Press.

Ayala, J., & Arp, R. (Eds.). (2010). *Contemporary debates in philosophy of biology*. Oxford: Wiley-Blackwell.

Ayala, F. J., & Dobzhansky, T. (Eds.). (1974). *Studies in the philosophy of biology*. Berkeley: University of California Press.

Baker, S. G., & Kramer, B. S. (2007). Paradoxes in carcinogenesis: New opportunities for research directions. *BMC Cancer, 7*, 151.

M. Bertolaso, *Philosophy of Cancer*, History, Philosophy and Theory of the Life Sciences 18, DOI 10.1007/978-94-024-0865-2

Baker, S. G., Soto, A. M., Sonnenschein, C., Capuccino, A., Potter, J. D., & Kramer, B. S. (2009). Plausibility of stromal initiation of epithelial cancers without a mutation in the epithelium: A computer simulation of morphostats. *BMC Cancer, 9*, 89.

Baylin, S. B. (2005). DNA methylation and gene silencing in cancer. *Nature Clinical Practice Oncology, 2*, 4–11.

Bechtel, W. (2006). *Discovering cell mechanisms: The creation of modern cell biology*. Cambridge: Cambridge University Press.

Bechtel, W., & Richardson, C. (2010). *Discovering complexity. Decomposition and localization as strategies in scientific research*. Cambridge, MA/London: MIT Press.

Ben-Jonathan, N., et al. (1996). Extrapituitary prolactin: Distribution, regulation, functions, and clinical aspects. *Endocrine Reviews, 17*(6), 639–669.

Bergers, G., Hanahan, D., & Coussens, L. M. (1998). Angiogenesis of protein kinase B/Akt. *Current Opinion in Cell Biology, 10*, 262–267.

Berrier, A. L., & Yamada, K. M. (2007). Cell-matrix adhesion. *Journal of Cellular Physiology, 213*(3), 565–573.

Bertolaso, M. (2009a). The neoplastic process and the problems with the attribution of function. *Rivista di Biologia/Biology Forum, 102*, 273–296.

Bertolaso, M. (2009b). Towards an integrated view of the neoplastic phenomena in cancer research. *History and Philosophy of the Life Sciencies, 31*, 79–98.

Bertolaso, M. (2011a). Hierarchies and causal relationships in the interpretative models of cancer in special issue causation and disease in the post-genomic era. *History and Philosophy of the Life Science, 33*, 389–396.

Bertolaso, M. (2011b). The two sides of the hourglass: analytic and synthetic approaches in cancer research. *Ludus Vitalis, 19*, 73–95.

Bertolaso, M. (2012a). *Il Cancro come questione. Modelli interpretativi e presupposti epistemologici*. Milano: Franco Angeli.

Bertolaso, M. (2012b). La dimensione non riduzionista del riduzionismo nella ricerca sperimentale. Dai modelli molecolari a quelli sistemici nella ricerca sul cancro. *Rivista Di Filosofia Neo-Scolastica, 4*, 687–705.

Bertolaso, M. (2013a). *How science works: Choosing levels of explanation in biological sciences*. Roma: Aracne.

Bertolaso, M. (2013b). On the structure of biological explanations: Beyond functional ascriptions in cancer research. *Epistemologia, 36*, 112–130.

Bertolaso, M. (2013c). Mechanisms & biological mechanisms. In *New Catholic encyclopedia supplement 2012–2013: Ethics and philosophy*. Detroit: Robert L. Fastiggi.

Bertolaso M. (2013d). Indeterminación Biológica y las Perspectivas Sistémicas de la Biología Contemporáne – Biological Uncertainty and the Systemic Perspectives of Contemporary Biology. *Anuario Filosofico, 46*, 365–386.

Bertolaso M. (2013e). Sulla 'irriducibilità' della prospettiva sistemica in biologia. Un'analisi delle convergenze dei modelli interpretativi del cancro. In L. Urbani Ulivi (Ed.), *Strutture di Mondo* (Vol. II, pp. 143–169). Bologna: Il Mulino.

Bertolaso, M., & Caianiello, S. (2016). Robustness as organized heterogeneity. *Rivista di Filosofia Neo-Scolastica, CVIII*.

Bertolaso, M., Giuliani, A., & Filippi, S. (2013). The mesoscopic level and its epistemological relevance in systems biology. In *Recent advances in systems biology* (Vol. 1, pp. 1–18). New York: Nova Science Publishers, Inc.

Biava, P. M. (1999). Complexity and cancer. *Leadership* Medica 1. Vedi (accesso di marzo 2008).

Biava, P. M. (2002). *Complessità e biologia. Il cancro come patologia della comunicazione*. Milano: Mondadori.

Bigelow, J., & Pargetter, R. (1987). Functions. *Journal of Philosophy, 84*(4), 181–196.

Bishop, J. M., & Weinberg, R. A. (1996). *Molecular oncology in cancer medicine*. New York: Scientific American, Inc.

Biskind, M. S., & Biskind, G. S. (1944). Development of tumors in the rat ovary after transplantation in the spleen. *Proceedings of the Society for Experimental Biology and Medicine, 55*, 176–181.

Bissell, M. J., & Inman, J. (2008). Reprogramming stem cells is a microenvironmental task. *Proceedings of the National Academy of Sciences of the United States of America, 105*, 15637–15638.
Bissell, M. J., & Radisky, D. (2001). Putting tumours in context. *Nature Reviews Cancer, 1*, 46–54.
Bissell, M. J., Hall, H. G., & Parry, G. (1982). How does the extracellular matrix direct gene expression? *Journal of Theoretical Biology, 99*(1), 31–68.
Bissell, M. J., Kenny, P. A., & Radisky, D. C. (2005). Microenvironmental regulators of tissue structure and function also regulate tumor induction and progression: The role of extracellular matrix and its degrading enzymes. *Cold Spring Harbor Symposia on Quantitative Biology, 70*, 343–356.
Bissell, M. J., Rizki, A., & Mian, I. S. (2003). Tissue architecture: The ultimate regulator of breast epithelial function. *Current Opinion in Cell Biology, 15*(6), 753–762.
Bizzarri, M., & Cucina, A. (2007). Reversibilità ed irreversibilità della trasformazione neoplastica: ipotesi per un nuovo paradigma. In *Cancro, complessità e derive psicoanalitiche*. A cura di Franchi, F. Milano: Franco Angeli Editore.
Bizzarri, M., et al. (2008). Beyond the oncogene paradigm: Understanding complexity in cancerogenesis. *Acta Biotheoretica, 56*(3), 173–196.
Bizzarri, M., Giuliani, A., Cucina, A., D'Anselmi, F., Soto, A. M., & Sonnenschein, C. (2011). Fractal analysis in a systems biology approach to cancer. *Seminars in Cancer Biology, 21*, 175–182.
Blanpain, C. (2013). Tracing the cellular origin of cancer. *Nature Cell Biology, 15*, 126–134.
Blau, H. M., Brazelton, T. R., & Weimann, J. M. (2001). The evolving concept of a stem cell: Entity or function? *Cell, 105*(7), 829–841.
Boman, B. M., & Wicha, M. S. (2008). Cancer stem cells: A step toward the cure. *Journal of Clinical Oncology, 26*, 2795–2799.
Boogerd, F. C., Bruggeman, F. J., Richardson, R. C., Stephan, A., & Westerhoff, H. V. (2005). Emergence and its place in nature: A case study of biochemical networks. *Synthese, 145*, 131–164.
Boogerd, F. C., Bruggeman, F. J., Hofmeyr, J. H. S., & Westerhoff, H. V. (Eds.). (2007). *Systems biology. Philosophical foundations*. Amsterdam: Elsevier.
Boorse, C. (1976). What a theory of mental health should be. *Journal for the Theory of Social Behavior, 6*(1), 61–84.
Boorse, C. (1977). Health as a theoretical concept. *Philosophy of Science, 44*(4), 542–573.
Boorse, C. (1997). A rebuttal on the Halth. In J. M. Humber & R. F. Almeder (Eds.), *What is disease?* (pp. 1–134). Totowa: Humana Press.
Boorse, C. (2002). A rebuttal on functions. In A. Ariew, R. Cummins, M. Perlman (Eds.) (pp. 63–112). New York: Oxford.
Boutwell, R. K. (1978). Biochemical mechanism of tumour promotion. In *Carcinogenesis*, Vol. 2 Mechanisms of tumour promotion and cocarcinogenesis. New York: Raven Press.
Boveri, T. (1914). *Concerning the origin of malignant tumours*. Questa monografia fu pubblicata in Germania nel 1914 e tradotta nel 1929 in inglese dalla moglie dell'autore. Nel 2007 è stata pubblicata in un'edizione critica da H. Harris (cfr. voce bibliografica corrispondente).
Brabletz, T., Jung, A., Spaderna, S., Hlubek, F., & Kirchner, T. (2005). Migrating cancer stem cells: An integrated concept of malignant tumour progression. *Nature Reviews Cancer, 5*, 744–749.
Braillard, P. A. (2010). Systems biology and the mechanistic framework. *Studies in History and Philosophy of Biological and Biomedical Sciences, 32*, 43–62.
Brenner, S. (1998). Biological computation. In G. R. Bock & J. A. Goode (Eds.), *The limits of reductionism in biology, Novartis Foundation Symposium 213* (pp. 106–111). Chichester: Wiley.
Brigandt, I. (2006). Philosophical issues in experimental biology. *Biology and Philosophy, 21*, 423–435.
Brigandt, I., & Love, A. (2012). *Reductionism in biology*. Stanford: Stanford Encyclopedia of Philosophy.
Bruce, W. R., & Van Der Gaag, H. (1963). A quantitative assay for the number of murine lymphoma cells capable of proliferation in vivo. *Nature, 199*, 79–80.

Bruggeman, F. J., Westerhoff, H. V., & Boogerd, F. C. (2002). BioComplexity: A pluralist research strategy is necessary for a mechanistic explanation of the 'live' state. *Philosophical Pshycology, 15*, 411–440.
Bunge, M. (1979). *Causality and modern science* (Vol. 7). New York: Dover Publications.
Bunge, M. (2003). *Emergence and convergence*. Toronto: University of Toronto Press.
Burmeister, T. (2001). Oncogenic retroviruses in animals and humans. *Reviews in Medical Virology, 11*, 369–380.
Buzzoni, M. (1995). *Scienza e Tecnica. Teoria ed esperienza nelle scienze della natura*. Roma: Studium.
Buzzoni, M. (2003). On medicine as a human science. *Theoretical Medicine, 24*, 79–94.
Buzzoni, M. (2015). Causality, teleology, and thought experiments in biology. *Journal of General Philosophy of Science*, *46*, 279–299.
Cairns, J. (1975). Mutation selection and the natural history of cancer. *Nature, 255*, 197–200.
Calin, G. A., et al. (2003). Genetic chaos and antichaos in human cancers. *Medical Hypotheses, 60*(2), 258–262.
Calotta, F. (2008). *Darwin contro il cancro*. Roma: Giovanni Fioriti Editore.
Campisi, J. (2005). Aging tumor suppression and cancer: A high wire act. *Mechanisms of Ageing and Development, 26*, 51–58.
Capp, J. P. (2005). Stochastic gene expression, disruption of tissue averaging effects, and cancer as a disease of development. *Bioessays, 27*, 1277–1285.
Cartwright, N. D. (1986). Two kind of teleological explanation. In A. Donagan, A. N. Perovich, Jr., & M. V. Wedin (Eds.), *Human nature and natural knowledge* (pp. 201–210). Dordrecht: Reidel.
Cavenee, W. K., Dryja, T. P., Phillips, R. A., Benedict, W. F., Godbout, R., Gallie, B. L., Murphreeparallel, A. L., Strong, L. C., & White, R. L. (1983). Expression of recessive alleles by chromosomal mechanisms in retinoblastoma. *Nature, 305*, 779–784.
Challis, G. B., & Stam, H. J. (1990). The spontaneous regression of cancer. A review of cases from 1900 to 1987. *Acta Oncologica, 29*, 545–550.
Cherubini, C., Gizzi, A., Bertolaso, M., Tambone, V., & Filippi, S. (2011). A bistable field model of cancer dynamics. *Communications in Computational Physics, 11*, 1–18.
Cicalese, A., et al. (2009). The tumor suppressor p53 regulates polarity of self-renewing divisions in mammary stem cells. *Cell, 138*(6), 1083–1095.
Clarke, M. F., Dick, J. E., Dirks, P. B., Eaves, C. J., Jamieson, C. H. M., Jones, D. L., Visvader, J., Weissman, I. L., & Wahl, G. M. (2006). Cancer stem cells—Perspectives on current status and future directions: AACR workshop on cancer stem cells. *Cancer Research, 66*, 9339–9344.
Cleaver, J. E. (2005). Cancer in xeroderma pigmentosum and related disorders of DNA repair. *Nature Reviews Cancer, 5*, 564–573.
Coffey, D. S. (1998). Self-organization, complexity and chaos: The new biology for medicine. *Nature Medicine, 4*(8), 882–885.
Coffman, J. A. (2011). On causality in nonlinear complex systems: The developmentalist perspective. In C. Hooker (Ed.), *Philosophy of complex systems*. Oxford/Waltham: North Holland.
Colditz, G. A., Sellers, T. A., & Trapido, E. (2006). Epidemiology–identifying the causes and preventability of cancer? *Nature Reviews Cancer, 6*, 75–83.
Coleman, W. B., et al. (1993). Regulation of the differentiation of diploid and some aneuploid rat liver epithelial (stemlike) cells by the hepatic microenvironment. *The American Journal of Pathology, 142*(5), 1373–1382.
Collins, F. S., & Barker, A. D. (2007). Mapping the cancer genome. Pinpointing the genes involved in cancer will help chart a new course across the complex landscape of human malignancies. *Scientific American, 296*, 50–57.
Condeelis, J., & Pollard, J. W. (2006). Macrophages: Obligate partners for tumor cell migration, invasion, and metastasis. *Cell, 124*, 263–266.
Cooper, G. M., Okenquist, S., & Siiverman, L. (1980). Transforming activity of DNA of chemically transformed and normal cells. *Nature, 284*, 415–421.
Craver, C. F. (2001). Role functions, mechanisms, and hierarchy. *Philosophy of Science, 68*, 53–74.

Craver, C. F. (2007). *Explaining the brain. Introduction*. Oxford: Claredon Press.
Craver, C. (2009). Mechanisms and natural kinds. *Philosophical Psychology, 22*, 575–594.
Cummins, R. (1975). Functional analysis. *The Journal of Philosophy, 72*, 741–765.
Dalerba, P., Cho, R. W., & Clarke, M. F. (2007). Cancer stem cells: Models and concepts. *Annual Review of Medicine, 58*, 267–284.
Darden, L., & Craver, K. (2009). *Reductionism in biology. Encyclopedia of life sciences* (pp. 1–6). Hoboken: Wiley.
Dawkins, R. (1976). *The selfish gene*. Oxford: Oxford University Press.
De Vita, V. T., Lawrence, T. S., & Rosenberg, S. A. (2008). *Cancer. Principles & practice of oncology* (Fig. 9.3). Philadelphia: Lippincott Williams & Wilkins.
Delehanty, M. (2005). Emergent properties and the context objection to reduction. *Biology and Philosophy, 20*, 715–734.
Dennett, D. (1996). *Darwin's dangerous idea*. New York: Touchstone.
Dexter, D. L., et al. (1978). Heterogeneity of tumor cells from a single mouse mammary tumor. *Cancer Research, 38*(10), 3174–3181.
Dobhzansky, T. H. (1973). Nothing in biology makes sense except in the light of evolution. *American Biology Teacher, 35*, 125–129.
Domazet-Loso, T., & Tautz, D. (2010). Phylostratigraphic tracking of cancer genes suggests a link to the emergence of multicellularity in metazoan. *BMC Biology, 21*, 8–66.
Duesberg, P. H. (1980). Transforming genes of retroviruses. *Cold Spring Harbor Symposium on Quantitative Biology, 44*, 13–29.
Dupré, J. (2003). *Promises and limits of reductionism in the biomedical sciences* (M. H. V. Van Regenmortel & D. L. Hull, Eds.). Hoboken: Wiley.
Dupré, J. (2010). *It is not possible to reduce biological explanations to explanations in chemistry and / or physics*. In J. Ayala & R. Arp (Eds.), *Contemporary debate in philosophy of biology*. Oxford: Wiley-Blackwell.
Dupré, J. (2012). *Processes of life*. Oxford: Oxford University Press.
Edelman, E. J., Guinney, J., Chi, J. T., Febbo, P. G., & Mukherijee, S. (2008). Modeling cancer progression via pathway dependencies. *PLoS Computational Biology, 4*, 21–13.
Eden, A., Gaudet, F., Waghmare, A., & Jaenisch, R. (2003). Chromosomal instability and tumors promoted by DNA hypomethylation. *Science, 300*, 455.
Editorial, N. (2006). All systems go! In *System Biology: A user's guide. Nature Reviews Molecular Cell Biology, 1179*.
Egger, G., Liang, G., Aparicio, A., & Jones, P. A. (2004). Epigenetics in human disease and prospects for epigenetic therapy. *Nature, 429*, 457–463.
Ereshefsky, M. (2009). Defining 'health' and 'disease'. *Studies in History and Philosophy of Biological and Biomedical Sciences, 40*, 221–227.
Farge, E. (2003). Mechanical induction of twist in the Drosophila foregut/stomodeal primordium. *Current Biology, 13*, 1365–1377.
Fearon, E. R., & Vogelstein, B. (1990). Genetic model for coloreotal tumorigenesis. *Cell, 61*, 759–767.
Feinberg, A. P. (2007). Phenotypic plasticity and the epigenetics of human disease. *Nature, 447*, 433–440.
Feinberg, A. P., & Tycko, B. (2004). The history of cancer epigenetics. *Nature Reviews Cancer, 4*, 143–153.
Feinberg, A. P., & Vogelstein, B. (1983). Hypomethylation distinguishes genes of some human cancers from their normal counterparts. *Nature, 301*, 89–92.
Feinberg, A. P., Ohlsson, R., & Henikoff, S. (2006). The epigenetic progenitor origin of human cancer. *Nature Reviews Genetics, 7*, 21–33.
Felli, N., et al. (2010). Hematopoietic differentiation: A coordinated dynamical process towards attractor stable states. *BMC Systems Biology, 4*, 85.
Ferrater, M. J. (1994). *Diccionario de Filiosofia*. Bercelona: Ariel.
Fidler, I. J. (2003). The pathogenesis of cancer metastasis: The "seed and soil" hypothesis revisited. *Nature Reviews Cancer, 3*, 453–458.

Fidler, I. J., & Kripke, M. L. (1977). Metastasis results from preexisting variant cells within a malignant tumor. *Science, 197*, 893–895.
Finkel, T., Serrano, M., & Blasco, M. A. (2007). The common biology of cancer and ageing. *Nature, 448*, 767–774.
Fogarty, M. P., Kessler, J. D., & Wechsler-Reya, R. J. (2005). Morphing into cancer: The role of developmental signalling pathways in brain tumor formation. *Journal of Neurobiology, 64*, 458–475.
Foulds, L. (1954). The experimental study of tumor progression: A review. *Cancer Research, 14*, 327–399.
Fox Keller, E. (2004). Marrying the premodern to the postmodern: Computers and organisms after World War II. In M. Norton (Ed.), *Wise growing explanations. Historical perspectives on recent science*. Durham: Duke University Press.
Fox Keller, E. (2010). It is possible to reduce biological explanations to explanations in chemistry and/or physiscs? In J. Ayala & R. Arp (Eds.), Contemporary debates in philosophy of biology. Malden/Chichester: Wiley-Blackwell.
Frank, S. (2003). Somatic mutation: Early cancer steps depend on tissue architecture. *Current Biology, 13*, 261–263.
Friend, S. H., Bernards, R., Rogelj, S., Weinberg, R. A., Rapaport, J. M., Albert, D. M., & Dryja, T. P. (1986). A human DMA segment with properties of the gene that predisposes to retinoblastoma and osteosarcoma. *Nature, 323*, 643–646.
Frisch, S. M., & Screaton, R. A. (2001). Anoikis mechanisms. *Current Opinion in Cell Biology, 13*, 555–562.
Fukushima, S., Kinoshita, A., Puatanachokchai, R., Kushida, M., Wanibuchi, H., & Morimura, K. (2005). Hormesis and dose-response-mediated mechanisms in carcinogenesis: Evidence for a threshold in carcinogenicity of non-genotoxic carcinogens. *Carcinogenesis, 26*, 1835–1845.
Fung, Y. K., Murphree, A. L., T'Ang, A., Qian, J., Hinrichs, S. H., & Benedict, W. F. (1987). Structural evidence for the authenticity of the human retinoblastoma gene. *Science, 236*, 1657–1661.
Fusco, G., Carrer, R., & Serrelli, E. (2014). The landscape metaphor in development. In A. Minelli & T. Pradeu (Eds.), *Towards a theory of development* (pp. 114–128). Oxford: Oxford University Press. doi:10.1093/acprof:oso/9780199671427.003.0007.
Gayon, J. (2006). Les biologistes ont-ils besoin du concept de fonction ? Perspective philosophique. *Comptes Rendus de l'Académie des Sciences, série Palevol.*
Ge, H., Walhout, J. M., & Vidal, M. (2003). Integrating 'omic' information: A bridge between genomics and systems biology. *Trends in Genetics, 19*, 551–560.
Germain, P. L. (2012). Cancer cells and adaptive explanations. *Biological Philosophy, 27*, 785–810.
Gilbert, S. F. (2005). Mechanisms for the environmental regulation of gene expression: Ecological aspects of animal development. *Journal of Biosciences, 30*, 65–74.
Gilbert, S. F., & Jorgensen, E. M. (1998). Wormwholes: A commentary on Schaffner, K.F. "Genes, Behaviour, and Developmental Emergentism". *Philosophy of Science, 65*, 259–266.
Gilbert, S. F., & Sarkar, S. (2000). Embracing complexity: Organicism for the 21st century. *Developmental Dynamics, 219*, 1–9.
Giroux, E. (2009). Définir objectivement la santé: une évaluation du concept bio-statistique de Boorse à partir de l'épidémiologie moderne. *Revue Philosophique, 1*, 35–58.
Giuliani, A. (2010). Collective motions and specific effectors: A statistical mechanics perspective on biological regulation. *BMC Genomics, 11*(Suppl 1), S2.
Giuliani, A., & Zbilut, J. P. (2009). *L'ordine della complessità*. Milano: Jaka Book.
Greaves, M. (2001). *Cancer: The evolutionary legacy*. Oxford: Oxford University Press.
Greaves, M., & Maley, C. C. (2012). Clonal evolution in cancer. *Nature, 481*, 306–313.
Green, D. R. (1999). Apoptotic pathways: The roads to ruin. *Cell, 94*, 695–698.
Greenman, C., Stephens, P., Smith, R., Dalgliesh, G. L., Hunter, C., Bignell, G., Davies, H., Teague, J., Butler, A., Stevens, C., Edkins, S., O'Meara, S., Vastrik, I., Schmidt, E. E., Avis, T., Barthorpe, S., Bhamra, G., Buck, G., Choudhury, B., Clements, J., Cole, J., Dicks, E., Forbes,

S., Gray, K., Halliday, K., Harrison, R., Hills, K., Hinton, J., Jenkinson, A., Jones, D., Menzies, A., Mironenko, T., Perry, J., Raine, K., Richardson, D., Shepherd, R., Small, A., Tofts, C., Varian, J., Webb, T., West, S., Widaa, S., Yates, A., Cahill, D. P., Louis, D. N., Goldstraw, P., Nicholson, A. G., Brasseur, F., Looijenga, L., Weber, B. L., Chiew, Y. E., De Fazio, A., Greaves, M. F., Green, A. R., Campbell, P., Birney, E., Easton, D. F., Chenevix- Trench, G., Tan, M. H., Khoo, S. K., Teh, B. T., Yuen, S. T., Leung, S. Y., Wooster, R., Futreal, P. A., & Stratton, M. R. (2007). Patterns of somatic mutation in human cancer genome. *Nature, 446*, 153–158.

Greger, V., Passarge, E., Hopping, W., Messmer, E., & Horsthemke, B. (1989). Epigenetic changes may contribute to the formation and spontaneous regression of retinobiastoma. *Human Genetics, 83*, 155–158.

Greten, F. R., Eckmann, L., Greten, T. F., Park, J. M., Li, Z. W., Egan, L. J., Kagnoff, M. F., & Karin, M. (2004). IKKbeta links inflammation and tumorigenesis in a mouse model of colitis-associated cancer. *Cell, 118*, 285–296.

Grier, D. G., Thompson, A., Kwasniewska, A., Mc-Gonigle, G. J., Halliday, H. L., & Lappin, T. R. (2005). The pathophysiology of HOX genes and their role in cancer. *Journal of Pathology, 205*(2), 154–171.

Griffiths, P. E., & Knight, R. D. (1998). What is the developmentalist challenge? *Philosophy of Science, 65*, 253–258.

Gupta, P. B., Chaffer, C. L., & Weinberg, R. A. (2009). Cancer stem cells: Mirage or reality? *Nature Medicine, 15*, 1010–1012.

Hagios, C., Lochter, A., & Bissell, M. J. (1998). Tissue architecture: The ultimate regulator of epithelial function? *Philosophical Transactions of the Royal Society of London. Series B, Biological Sciences, 353*(1370), 857–870.

Hahn, W. C., & Weinberg, R. A. (2002a). Modelling the molecular circuitry of cancer. *Nature Reviews Cancer, 2*, 331–342.

Hahn, W. C., & Weinberg, R. A. (2002b). Rules for making human tumor cells. *New England Journal of Medicine, 347*, 1593–1604.

Hahn, W. C., Counter, C. M., Lundberg, A. S., Beijersbgern, R. L., Brooks, M. V., & Weinberg, R. A. (1999). Creation of human tumor cells with defined genetic elements. *Nature, 400*, 464–468.

Hamburger, A., & Salmon, S. E. (1977). Primary bioassay of human myeloma stem cells. *The Journal of Clinical Investigation, 60*(4), 846–854.

Hanahan, D., & Weinberg, R. A. (2000). The hallmarks of cancer. *Cell, 100*(1), 57–70.

Hanahan, D., & Weinberg, R. A. (2011). Hallmarks of cancer: The next generation. *Cell, 144*, 646–674.

Harris, M. (1971). Cell fusion and the analysis of malignancy. *Proceedings of the Royal Society of London: Biology Science, 179*, 1–20.

Harris, H. (1986). Cancer: Malignant tumours generated by recessive mutations. *Nature, 323*(6089), 582–583.

Harris, H. (2003). Is collagen XV a tumor suppressor? *DNA and Cell Biology, 22*, 225–226.

Harris, H. (2004). Tumour suppression: Putting on the brakes. *Nature, 427*, 201.

Harris, H. (2005). A long view of fashions in cancer research. *BioEssays: News and Reviews in Molecular, Cellular and Developmental Biology, 27*(8), 833–838.

Harris, A., Harris, H., & Hollingsworth, M. A. (2007). Complete suppression of tumor formation by high levels of basement membrane collagen. *Molecular Cancer Research: MCR, 5*(12), 1241–1245.

Harris, H., et al. (1969). Suppression of malignancy by cell fusion. *Nature, 223*(5204), 363–368.

Harris, H., Rawlins, J., & Sharps, J. (1996). A different approach to tumour suppression. *Journal of Cell Science, 109*, 2189–2197.

Hendrix, M. J., Seftor, E. A., Seftor, R. E., Kasemeier-Kulesa, J., Kulesa, P. M., & Postovit, L. M. (2007). Reprogramming metastatic tumour cells with embryonic microenvironments. *Nature Reviews Cancer, 7*, 246–255.

Heng, H. H., Bremer, S. W., Stevens, J., Ye, K. J., Miller, F., Liu, G., & Ye, C. J. (2006a). Cancer progression by non-clonal chromosome aberrations. *Journal of Cellular Biochemistry, 98*, 1424–1435.

Heng, H. H., Liu, G., Bremer, S., Ye, K. J., Stevens, J., & Ye, C. J. (2006b). Clonal and non-clonal chromosome aberrations and genome variation and aberration. *Genome, 49*, 195–204.

Heng, H. H., Stevens, J. B., Liu, G., Bremer, S. W., Ye, K. J., Reddy, P. V., Wu, G. S., Wang, Y. A., Tainsky, M. A., & Ye, C. J. (2006c). Stochastic cancer progression driven by non-clonal chromosome aberrations. *Journal of Cellular Physiology, 208*, 461–472.

Heng, H. H., Stevens, J. B., Lawrenson, L., Liu, G., Ye, K. J., Bremer, S. W., & Ye, C. J. (2008). Patterns of genome dynamics and cancer evolution. *Cellular Oncology, 30*, 513–514.

Heng, H. H., Bremer, S. W., Stevens, J. B., Ye, K. J., Liu, G., & Ye, C. J. (2009). Genetic and epigenetic heterogeneity in cancer: A genome-centric perspective. *Journal of Cellular Physiology, 220*, 538–547.

Heppner, G. H. (1984). Tumor heterogeneity. *Cancer Research, 44*, 2259–2265.

Heppner, G. H., & Miller, F. R. (1998). The cellular basis of tumor progression. *International Review of Cytology, 177*, 1–56.

Hetzer, M. W., Walther, T. C., & Mattaj, I. W. (2005). Pushing the envelope: Structure, function, and dynamics of the nuclear periphery. *Annual Review of Cell and Developmental Biology, 21*, 347–380.

Hillenmeyer, M. E., Fung, E., Wildenhain, J., Pierce, S. E., Hoon, S., Lee, W., Proctor, M., St Onge, R. P., Tyers, M., Koller, D., Altman, R. B., Davis, R. W., Nislow, C., & Giaever, G. (2008). The chemical genomic portrait of yeast: Uncovering a phenotype for all genes. *Science, 320*, 362–365.

Hochedlinger, K., Blelloch, R., Brennan, C., Yamada, Y., Kim, M., Chin, L., & Jaenisch, R. (2004). Reprogramming of a melanoma genome by nuclear transplantation. *Genes & Development, 18*, 1875–1885.

Holland, J. F. 2003. *Cancer Medicine*, Part I. London: BC Decker Inc Hamilton.

Hornberg, J. J., Bruggeman, F. J., Westerhoff, H. W., & Lankelma, J. (2006). Cancer: A systems biology disease. *BioSystems, 83*, 81–90.

Hough, C. D., et al. (1997). Organizing a functional junctional complex requires specific domains of the Drosophila MAGUK Discs large. *Genes & Development, 11*(23), 3242–3253.

Howlett, A. R., et al. (1995). Cellular growth and survival are mediated by beta 1 integrins in normal human breast epithelium but not in breast carcinoma. *Journal of Cell Science, 108*(Pt 5), 1945–1957.

Huang, S., & Ingber, D. E. (2000). Shape-dependent control of cell growth, differentiation, and apoptosis: Switching between attractors in cell regulatory networks. *Experimental Cell Research, 261*, 91–103.

Huang, S., & Ingber, D. E. (2007). A non-genetic basis for cancer progression and metastasis: Self-organizing attractors in cell regulatory networks. *Breast Disease, 26*, 27–54.

Huang, S., & Wikswo, J. (2006). Dimensions of systems biology. *Reviews of Physiology, Biochemistry and Pharmacology, 157*, 81–104.

Huang, H. J., Yee, J. K., Shew, J. Y., Chen, P. L., Bookstein, R., Friedmann, T., Lee, E. Y., & Lee, W. H. (1988). Suppression of the neoplastic phenotype by replacement of the RB gene in human cancer cells. *Science, 242*, 1563–1566.

Huang, A., et al. (2002). Serum tryptophan decrease correlates with immune activation and impaired quality of life in colorectal cancer. *British Journal of Cancer, 86*(11), 1691–1696.

Huang, S., Ernbergb, I., & Kauffman, S. (2009). Cancer attractors: A systems view of tumours from a gene network dynamics and developmental perspective. *Seminars in Cell & Developmental Biology, 20*, 869–887.

Hull, D. L. (1976). *Informal aspects of theory reduction.* In R. S. Cohen, C. A. Hooker, C. A. Michalos, J. W. Evra, D. PSA, & D. Reidel (Eds.). 1974. *Boston studies in the philosophy of science* (pp. 653–670).

Hull, D. L. (1981). Philosophy and biology. In G. Eløistad (Ed.) *Contemporary philosophy: A new survey* (Vol. 2, pp. 281–316). The Hague: Martinus Nijhoff.

Hull, D. L. (2002). Recent philosophy of biology: A review. *Acta Biotheoretica, 50*, 117–128.

Hüttemann, A., & Love, A. C. (2011). Aspects of reductive explanation in biological science: Intrinsicality, fundamentality and temporality. *The British Journal for the Philosophy of Science, 62*, 519–549.

Ingber, D. E. (2000). The origin of cellular life. *BioEssays: News and Reviews in Molecular, Cellular and Developmental Biology, 22*(12), 1160–1170.

Ingber, D. E. (2008). Can cancer be reversed by engineering the tumor microenvironment? *Seminars in Cancer Biology, 18*, 356–364.

Istituti Fisioterapici Ospitalieri (Ifo). Vedi http://www.ifo.it/AspOne.aspx?990000242. accesso del 16 agosto 2008.

Jacks, T., & Weinberg, R. A. (2002). Taking the study of cancer cell survival to a new dimension. *Cell, 111*, 923–925.

Jaffe, L. (2005). Response to paper by Henry Harris. *Bioessays, 27*, 1206.

Jones, P. A., & Baylin, S. B. (2007). The epigenomics of cancer. *Cell, 128*, 683–692.

Jones, P. A., Archer, T.K., Baylin, S.B., Beck, S., Berger, S., Bernstein, B. E., Carpten, J. D., Clark, S. J., Costello, J. F., Doerge, R. W., Esteller, M., Feinberg, A. P., Gingeras, T. R., Greally, J. M., Henikoff, S., Herman, J. G., Jackson-Grusby, L., Jenuwein, T., Jirtle, R. L., Kim, Y. J., Laird, P. W., Lim, B., Martienssen, R., Polyak, K., Stunnenberg, H., Tlsty, T. D., Tycko, B., Ushijima, T., Zhu, J., Pirrotta, V., Allis, C. D., Elgin, S. C., Jones, P. A., Martienssen, R., Rine, J., & Wu, C. (2008). American Association for Cancer Research Human Epigenome Task Force; European Union, Network of Excellence, Scientific Advisory Board. Moving AHEAD with an international human epigenome project. *Nature, 454*, 711–715.

Juarrero, A. (2002). *Dynamics in action. Intentional behaviour as a complex system*. Cambridge, MA: MIT Press.

Kahn, C. R., et al. (1978). Direct demonstration that receptor crosslinking or aggregation is important in insulin action. *Proceedings of the National Academy of Sciences of the United States of America, 75*(9), 4209–4213.

Kalluri, R. (2003). Basement membranes: Structure, assembly and role in tumour angiogenesis. *Nature Reviews Cancer, 3*, 422–433.

Katoh, M. (2008). Wnt10b. UCSD-Nature Molecule Pages. Vedi http://www.signalinggateway.org/molecule/query?afcsid=A002375. accesso del 10 giugno 2010.

Kauffman, S. A. (1970). Articulation of parts: Explanation in biology and the rational search for them. *Boston Studies in Philosophy of Science, 8*, 257–272.

Keller, E. F. (2010). It is possible to reduce biological explanations to explanations in chemistry and/or physiscs? In J. Ayala & R. Arp (Eds.), *Contemporary debates in philosophy of biology*. Oxford: Wiley-Blackwell.

Kelly, P. N., Dakic, A., Adams, J. M., Nutt, S. L., & Strasser, A. (2007). Tumor growth need not be driven by rare cancer stem cells. *Science, 317*, 337.

Kemler, R. (1993). From cadherins to catenins: Cytoplasmic protein interactions and regulation of cell adhesion. *Trends in Genetics, 9*(9), 317–321.

Kenny, P. A., & Bissell, M. J. (2003). Tumor reversion: Correction of malignant behaviour by microenvironmental cues. *International Journal of Cancer, 107*, 688–695.

Khalil, I. G., & Hill, C. (2005). Systems biology for cancer. *Current Opinion in Oncology, 17*, 44–48.

Kim, J. (1999). Making sense of emergence. *Philosophical Studies, 95*, 3–36.

Kingma, E. (2009). Paracetamol, poison, and polio: Why Boorse's account of function fails to distinguish health and disease. *The British Journal for Philosophy of Science, 61*, 1–24.

Kinzler, K. W., & Vogelstein, B. (1997). Cancer-susceptibility genes. Gatekeepers and caretakers. *Nature, 386*, 761–763.

Kinzler, K. W., & Vogelstein, B. (1998). Landscaping the cancer terrain. *Science (New York, NY), 280*(5366), 1036–1037.

Kirshner, J., Schumann, D., & Shively, J. E. (2003). CEACAM1, a cell-cell adhesion molecule, directly associates with annexin II in a three-dimensional model of mammary morphogenesis. *The Journal of Biological Chemistry, 278*(50), 50338–50345.

Kitano, H. (2002). Systems biology: A brief overview. *Science, 295*, 1662–1664.

Kitano, H. (2004a). Cancer as a robust system: Implication for anticancer therapy. *Nature Reviews Cancer*, 4, 227–235.

Kitano, H. (2004b). Biological robustness. *Nature Reviews Genetics, 5*(11), 826–837.

Kitcher, P. (1999). The egemony of molecular biology. *Biology and Philosophy, 14*, 195–210.

Klein, G. (2002). Perspectives in studies of human tumor viruses. *Frontiers in Bioscience, 7*, d268–d274.

Klein, G., & Klein, E. (1986). Conditioned tumorigenicity of activated oncogenes. *Cancer Research, 46*, 3211–3224.

Knudson, A. G. (1971). Mutation and cancer: Statistical study of retinoblastoma. *Proceedings of the National Academy of Sciences, 68*(4), 820–823.

Krohs, U., & Callebaut, W. (2007). In F. C. Boogerd, F. J. Bruggeman, J. H. S. Hofmeyer, & H. V. Westerhoff (Eds.), *Systems biology. Philosophical foundations* (pp. 181–214). Elsevier. op. cit.

Laforge, B., Guez, D., Martinez, M., & Kupiec, J. J. (2005). Modeling embryogenesis and cancer: An approach based on an equilibrium between the autostabilization of stochastic gene expression and the interdependence of cells for proliferation. *Progress in Biophysics and Molecular biology, 89*(1), 93–120.

Land, H., Parada, L. F., & Weinberg, R. A. (1983). Tumorigenic conversion of primary embryo fibroblasts requires at least two cooperating oncogenes. *Nature, 304*(5927), 596–602.

Lee, M. G., & Nurse, P. M. (1987). Complementation used to clone a human homologue of the fission yeast cell cycle control gene. *Nature, 327*, 31–35.

Lee, W. H., et al. (1987). Human retinoblastoma susceptibility gene: Cloning, identification, and sequence. *Science, 235*(4794), 1394–1399.

Leonelli, S. (2012). Making sense of data-driven research in the biological and the biomedical sciences. *Studies in the History and Philosophy of the Biological and Biomedical Sciences, 43*, 1–3.

Levins, R., & Lewontin, R. (1985). *The dialectical biologist*. Cambridge: Harvard University Press.

Lewis, D. (2000). Causation as influence. *The Journal of Philosophy, 47*, 182–197.

Lewontin, R. (1970). The units of selection. *Annual Review of Ecology and Systematics, 1*, 1–18.

Lewontin, R., & Levins, R. (2007). *Biology under the influence* (p. 105). New York: Monthly Review Press.

Licata, I. (2015). Incertezza. Un approccio sistemico. In L. Urbani Ulivi (Ed.), *Strutture di Mondo* (Vol. III, pp. 35–73 and 46–47). Il Mulino.

Lijinsky, W. (1989). A view of the relation between carcinogenesis and mutagenesis. *Environmental and Molecular Mutagenesis, 14*(Suppl), 1678–1684.

Liu, H., Radisky, D. C., Wang, F., & Bissell, M. J. (2004). Polarity and proliferation are controlled by distinct signaling pathways downstream of PI3-kinase in breast epithelial tumor cells. *Journal of Cell Biology, 164*, 603–612.

Lobo, N. A., Shimono, Y., Qian, D., & Clarke, M. F. (2007). The biology of cancer stem cells. *Annual Review of Cell Developmental Biology 23*, 675–699, Table 1.

Longtin, R. (2005). For tissue organization theory of cancer, a difficult road to acceptance. *JNCI Journal of the National Cancer Institute, 97*(1), 11–12.

Lotem J, Sachs L (1974, September) Different blocks in the differentiation of myeloid leukemic cells. *Proceedings of the National Academy of Sciences USA, 71*(9), 3507–3511.

Lotem, J., & Sachs, L. (2002). Epigenetics wins over genetics: Induction of differentiation in tumor cells. *Seminars in Cancer Biology, 12*, 339–346.

Lowe, S. W., Cepero, E., & Evan, G. (2004). Intrinsic tumor suppression. *Nature, 432*, 307–315.

Luch, A. (2006). The mode of action of organic carcinogens on cellular structures. *Experientia. Supplementum, 96*, 65–95.

Machamer, P. K., & Silberstein, M. (Eds.). (2002). Guide to the philosophy of science. *Blackwell.*

Machamer, P., Darden, L., & Craver, C. (2000). Thinking about mechanisms. *Philosophy of Science, 67*, 1–25.

Maffini, M. V., Soto, A. M., Calabro, J. M., Ucci, A. A., & Sonnenschein, C. (2004). The stroma as a crucial target in rat mammary gland carcinogenesis. *Journal of Cell Science, 117*, 1495–1502.

Magee, J. a., Piskounova, E., & Morrison, S. J. (2012). Cancer stem cells: Impact, heterogeneity, and uncertainty. *Cancer Cell, 21*(3), 283–296.

Maitra, A., Arking, D. E., Shivapurkar, N., Ikeda, M., Stastny, V., Kassauei, K., Sui, G., Cutler, D. J., Liu, Y., Brimble, S. N., Noaksson, K., Hyllner, J., Schulz, T. C., Zeng, X., Freed, W. J., Crook, J., Abraham, S., Colman, A., Sartipy, P., Matsui, S. I., Carpenter, M., Gazdar, A. F., Rao, M., & Chakravarti, A. (2005). Genomic alterations in cultured human embryonic stem cells. *Nature Genetics, 37*, 1099–1103.

Malaterre, C. (2007). Organicism and reductionism in cancer research: Towards a systemic approach. *International Studies in the Philosophy of Science, 21*(1), 57–73.

Malaterre, C. (2011). Making sense of downward causation in manipulationism: Illustrations from cancer research. *History and Philosophy of the Life Sciences, 33*, 537–561.

Malins, D. C., et al. (1998). A unified theory of carcinogenesis based on order – Disorder transitions in DNA structure as studied in the human ovary and breast. *Proceedings of the National Academy of Sciences of the United States of America, 95*(June), 7637–7642.

Marcos, A. (2009). Funciones en biologia: una perspective aristotélica. *Dialogo Filosófico, 74*, 231–248.

Marcos, A. (2012). *Postmodern Aristotele*. Newcastle: Cambridge Scholars Publishing.

Marcum, J. A. (2005). Metaphisical presuppositions and scientific practices: Reductionism and organicism in cancer research. *International Studies in the Philosophy of Science, 19*, 31–44.

Marker, P. C. (2008). Does prostate cancer co-opt the developmental program? *Differentiation, 76*, 736–744.

Marshall, C. J. (1991). Tumor suppressor genes. *Cell, 64*, 313–326.

Martien, S., & Abbadie, C. (2007). Acquisition of oxidative DNA damage during senescence: The first step toward carcinogenesis? *Annals of the New York Academy of Sciences, 1119*, 51–63.

Marte, B. (2009). Insights into the development of cancer. *Nature Collections, Cancer*. http://www.nature.com/nature/supplements/collections/cancer/editorial.html. Accessed 10 July 2016.

Mayr, E. (1982). *The growth of biological thought: Diversity, evolution, and inheritance*. Cambridge: Belknap Press.

Mayr, E. (2004). *L'unicità della Biologia. Sull'autonomia di una disciplina scientifica*. Raffaello Cortina Editore, 2005. (Ed., inglese: *What makes biology unique? Considerations on the autonomy of a scientific discipline*).

McCawley, L. J., Wright, J., LaFleur, B. J., Crawford, H. C., & Matrisian, L. M. (2008). Keratinocyte expression of MMP3 enhances differentiation and prevents tumor establishment. *American Journal of Pathology, 173*(5), 1528–1539.

McCullough, K. D., Coleman, W. B., Ricketts, S. L., Wilson, J. W., Smith, G. J., & Grisham, J. W. (1998). Plasticity of the neoplastic phenotype in vivo is regulated by epigenetic factors. *Proceedings of the National Academy of Sciences of the United States of America, 95*, 15333–15338.

McLauglin, P. (2001). *What functions explain. Functional explanation and self-reproducing systems*. Cambridge: Cambridge University Press.

Meadows, D. H., Meadows, D. L., & Randers, J. (1992). Beyond the limits. In R. Lewontin, & R. Levins (Eds.). (2007). *Biology under the influence* (p. 105). New York: Monthly Review Press.

Merlo, L. M., Pepper, J. W., Reid, B. J., & Maley, C. C. (2006). Cancer as an evolutionary and ecological process. *Nature Reviews Cancer, 6*, 924–935.

Michor, F., Iwasa, Y., & Nowak, M. A. (2004). Dynamics of cancer progression. *Nature Reviews Cancer, 4*, 197–205.

Mintz, B., & Illmensee, K. (1975). Normal genetically mosaic mice produced from malignant teratocarcinoma cells. *Proceedings of the National Academy of Sciences of the United States of America, 72*, 3585–3589.

Mitchell, S. D. (2000). Dimensions of scientific law. *Philosophy of Science, 67*, 242–265.

Mitchell, S. D. (2003). *Biological complexity and integrative pluralism*. Cambridge: Cambridge Studies in Philosophy and Biology.

Mitchell, S. D. (2005). Modularity: More than a Buzzword "Essay Review". *Biological Theory, 1*, 98–101.

Mitchell, S. D. (2008). Explaining complex behavior. In K. S. Kendler & J. Parnas (Eds.), *Philosophical issues in psychiatry: Explanation, phenomenology, and nosology* (pp. 19–38). Baltimore: Johns Hopkins University Press.

Mitchell, S. D. (2009). *Unsimple truths: Science, complexity and policy*. Chicago: University of Chicago Press.

Mitchell, S. D. (2012). Emergence: Logical, functional and dynamical. *Synthese, 185*, 171–186.

Monod. J. (1954). cit. in Weinberg, R.A. 2006. The biology of cancer. *Garland Science,* p. 1.

Morton, B. (1974). Reduction, hierarchies and organicism. In F. J Ayala & T. Dobzhansky (Eds.), *Studies in the philosophy of biology*. Berkeley: University of California Press.

Moss, L. (2002). *What genes can't do*. Cambridge: MIT Press.

Mossio, M., Saborido, C., & Moreno, A. (2009). An organizational account of biological functions. *The British Journal for Philosophy of Science, 60*(4), 813–841.

Murillo, J. I. (2010). El tiempo y los métodos de la biologia. *StudiaPoliana, 12*, 55–68.

Nagel, E. (1961). *The structure of science*. New York: Harcourt, Brace and World.

Nathanson, L. (1976). Spontaneous regression of malignant melanoma: A review of the literature on incidence, clinical features and possible mechanisms. *National Cancer Institute Monograph, 44*, 67–76.

National Cancer Institute. (1976). *Conference on spontaneous regression of cancer* (pp. 76–1038). Bethesda: US Department of Health, Education and Welfare, NIH publ.

Neander, K. (1991). Functions as selected effects : The conceptual analyst's defence. *Philosophy of Science, 5*, 169–184.

Neander, K. (2009). Les explications fonctionnelles. *Revue Philosophique, 1*, 5–34.

Nelson, C. M., & Bissell, M. J. (2005). Modeling dynamic reciprocity: Engineering three dimensional culture models of breast architecture, function, and neoplastic transformation. *Seminars in Cancer Biology, 15*, 342–352.

Nelson, C. M., & Bissell, M. J. (2006). Of extracellular matrix, scaffolds, and signaling: Tissue architecture regulates development, homeostasis, and cancer. *Annual Review of Cell and Developmental Biology, 22*, 287–309.

Neumann-Held, E., & Rehmann-Sutter, C. (Eds.). (2006). *Genes in development: Re-reading the molecular paradigm*. Durham: Duke University Press.

Nicholson, D. J. (2010). Biological atomism and cell theory. *Studies in History and Philosophy of Biological and Biomedical Sciences, 41*, 202–211.

Noble, D. (2003). The future: Putting Humpty Dumpty together again. *Biochemical Society Transactions, 31*, 156–158.

Noble, D. (2006). *The music of life: Biology beyond the genome*. Oxford: Oxford University Press.

Nowak, M. (2006). *Evolutionary dynamics: Exploring the equations of life*. Cambridge: Harvard University Press.

Nowell, P. C. (1976). The clonal evolution of tumor cell populations. *Science, 194*, 23–28.

Nowell, P. C., & Hungerford, D. A. (1960). A minute chromosome in human chronic granulocytic leukaemia. *Science, 132*, 1488–1501.

O'Malley, M. A., & Dupré, J. (2005). Fundamental issues in systems biology. *Bioessays, 27*, 1270–1276.

Oakley, E. J., & Van Zant, G. (2007). Unraveling the complex regulation of stem cells: Implications for aging and cancer. *Leukemia, 21*, 612–621.

Palumbo, M. C., et al. (2005). Functional essentiality from topology features in metabolic networks: A case study in yeast. *FEBS Letters, 579*(21), 4642–4646.

Palumbo, M. C., et al. (2007). Essentiality is an emergent property of metabolic network wiring. *FEBS Letters, 581*(13), 2485–2489.

Pardal, R., Clarke, M. F., & Morrison, S. J. (2003). Applying the principles of Stem-cell biology to cancer. *Nature Reviews Cancer, 3*, 895–902.

Parkin, D. M. (2004). International variation. *Oncogene, 23*, 6329–6340.

Penfield, P., Spence, R., & Duinker, S. (1970, August). Generalized form of Tellegen's theorem. *IEEE Transactions on Circuit Theory CT, 17*(3), 302–305.

Pepper, J. W., Sprouffske, K., & Maley, C. C. (2007). Animal cell differentiation patterns suppress somatic evolution. *PLoS Computational Biology, 3*(12), e250.

Perovic, S. (2007). The limitations of Kim's reductive physicalism in accounting for living systems and an alternative non reductionist ontology. *Acta Biotheoretica, 55*, 243–267.

Phelps, T. J., Palombo, A. V., & Beliaev, A. S. (2002). Metabolomics and microarrays for improved understanding of phenotypic characteristics controlled by both genomic and environmental constraints. *Current Opinion in Biotechnology, 13*, 20–24.

Pierce, G. B., Nakane, P. K., Martinez-Hernandez, A., & Ward, J. M. (1977). Ultrastructural comparison of differentiation of stem cells of murine adenocarcinomas of colon and breast with their normal counterparts. *Journal of the National Cancer Institute, 58*, 1329–1345.

Plutynski, A. (2013). Cancer and the goals of integration. *Studies in History and Philosophy of Biological and Biomedical Sciences, 44*(4), 466–476.

Posadas, E. M., Criley, S. R., & Coffey, D. S. (1996). Chaotic oscillations in cultured cells: Rat prostate cancer. *Cancer Research, 56*(16), 3682–3688.

Poste, G., & Greig, R. (1982). On the genesis and regulation of cellular heterogeneity in malignant tumors. *Invasion and Metastasis, 2*, 137–176.

Potochnik, A. (2010). Explanatory independence and epistemic interdependence: A case study of the optimality approach. *The British Journal for the Philosophy of Science, 61*, 213–233.

Potter, V. R. (1964). Biochemical perspective in cancer research. *Cancer Research, 24*, 1085–1098.

Potter, V. R. (1968). *Mechanisms of carcinogenesis in relation to studies on minimal deviation hepatomas*. In 1969. *Exploitable molecular mechanisms and neoplasia* (pp. 587–610). Austin: The University of Texas Press.

Potter, V. R. (1969). Recent trends in cancer biochemistry: The importance of studies on fetal tissues. *Proceedings of Cananadian Cancer Conference, 8*, 9–30.

Potter, V. R. (1978). Phenotypic diversity in experimental hepatomas: The concept of partially blocked ontogeny. The 10th Walter Hubert lecture. *British Journal of Cancer, 38*, 1–23.

Prehn, R. T. (1994). Cancers beget mutations versus mutations beget cancers. *Cancer Res, 54*, 5296–300. (Cfr. Van Speybroeck, L. (2002). From epigenesist to epigenetics: The case of CH Waddington. *Annals of New York Academic Science, 981*: 61–81).

Quintana, E., Shackleton, M., Sabel, M. S., Fullen, D. R., Johnson, T. M., & Morrison, S. J. (2008). Efficient tumour formation by single human melanoma cells. *Nature, 456*, 593–598.

Radisky, D. C., Levy, D. D., Littlepage, L. E., Liu, H., Nelson, C. M., Fata, J. E., Leake, D., Godden, E. L., Albertson, D. G., Nieto, M. A., Werb, Z., & Bissell, M. J. (2005). Rac1b and reactive oxygen species mediate MMP-3-induced EMT and genomic instability. *Nature, 436*, 123–127.

Rapp, U. R., Ceteci, F., & Schreck, R. (2008). Oncogene-induced plasticity and cancer stem cells. *Cell Cycle, 7*, 45–51.

Reya, T., Morrison, S. J., Clarke, M. F., & Weissman, I. L. (2001). Stem cells, cancer, and cancer stem cells. *Nature, 414*, 105–111.

Richardson, R. C., & Stephan, A. (2007). Mechanism and mechanical explanation in systems biology. In F. C. Boogerd, F. J. Bruggerman, J. S. Hofmeyr, & H. V. Westerhoff (Eds.), *Systems biology: Philosophical foundations* (pp. 123–144). New York: Elsevier.

Ritter, W. E., & Bailey, E. W. (1928). *The organismal conception: Its place in science and its bearing on philosophy*. Berkeley: University of California Press.

Root-Bernstein, R. S. (1999). Complementarity and contradiction in cancer research: The role of hierarchies in carcinogenesis. *Anticancer Research, 19*, 4915–4918.

Rosenberg, A. (2006). *Darwinian reductionism or how to stop wirrying and love molecular biology* (pp. 7–26). Chicago: University of Chicago Press.

Rosenberg, A. (2007). Reductionism (and anti reductionism) in biology. In D. L. Hull & M. Ruse (Eds.), *The Cambridge companion to the philosophy of biology.* Cambridge: Cambridge University Press.

Rosenberg, A., & Kaplan, D. M. (2005). How to reconcile physicalism and anti-reductionism about biology. *Philosophy in Science, 72*, 43–68.

Rosenberg, A., & McShea, D. (2008). *Philosophy of biology*. New York: Routledge.

Rous, P. (1910). A transmissible avian neoplasm (sarcoma of the common fowl). *Journal of Experimental Medicine, 12*, 696–705.

Rubin, H. (1999). Cell damage, aging and transformation: A multilevel analysis of carcinogenesis. *Anticancer Research, 19*(6A), 4877–4886.

Rubin, H. (2001). Fields and cancerization: The preneoplastic origins of cancer: Asymptomatic hyperplastic fields are precursors of neoplasia, and their progression to tumors can be tracked by saturation density in culture. *Bioessays, 33*, 224–231.

Rubin, H. (2007). Ordered heterogeneity and its decline in cancer and aging. *Advances in Cancer Research, 98*, 117–147.

Ruddon, R. W. (1995). *Cancer biology*. New York: Oxford University Press.

Ruse, M. (1976). Reduction in genetics. *Boston Studies in the Philosophy of Science, 32*, 631–651.

Saborido, C. (2012). *Funcionalidad y Organización en Biología*. Ph.D. dissertation, University of the Basque Country.

Saetzler, K., Sonnenschein, C., & Soto, A. M. (2011). Systems biology beyond networks: Generating order from disorder through self-organization. *Seminars in Cancer Biology, 21*(3), 165–174.

Scrable, H., et al. (1989). A model for embryonal rhabdomyosarcoma tumorigenesis that involves genome imprinting. *Proceedings of the National Academy of Sciences of the United States of America, 86*(19), 7480–7484.

Schaffner, K. F. (1976). Reductionism in biology: Prospects and problems. *Boston Studies in the Philosophy of Science, 32*, 613–632.

Schaffner, K. F. (1993). *Discovery and explanation in biology and medicine*. Chicago: University of Chicago Press.

Schaffner, K. F. (1998). Model organisms and behavioral genetics: A rejoinder. *Philosophy of Science, 65*, 276–288.

Schaffner, K. F. (2002). Neuroethics: Reductionism, emergence, and decision-making capacities. In S. J. Marcus (Ed.), *Neuroethics: Mapping the field* (pp. 27–33). New York: The Dana Foundation Press.

Schaffner, K. F. (2006). Reduction: The Cheshire Cat problem and a return to roots. *Synthese, 151*, 377–402.

Schaffner, K. F. (2007). Theories, models and equations in systems biology. In F. C. Boogerd, F. J. Bruggeman, J. H. S. Hofmeyer, & H. V. Westerhoff (Eds.), *Systems biology: Philosophical foundations* (p. 145). Amsterdam: Elsevier.

Schaffner, K. F. (2013a). *Behaving. What's genetic and what's not and why should we care? Monograph*. Oxford: Oxford University Press.

Schaffner, K. F. (2013b). Reduction and reductionism in psychiatry. In K. W. M. Fulford, M. Davies, R. G. T. Gipps, G. Graham, J. Z. Sadler, G. Stanghellini, & T. Thornton (Eds.), *The Oxford handbook of philosophy and psychiatry*. Oxford: Oxford University Press.

Schaffner, K. F. (2016). *Behaving: What's genetic, what's not, and why should we care?* Oxford: Oxford University Press.

Schlessinger, J. (1980). The mechanism and role of hormone-induced clustering of membrane receptors. *Trends in Biochemical Sciences, 5*(8), 210–214.

Schwartz, P. H. (2007). Defining dysfunction: Natural selection, design, and drawing a line. *Philosophy of Science, 74*, 364–385.

Sedivy, R. (1999). Chaodynamic loss of complexity and self-similarity in cancer. *Medical Hypotheses, 52*(4), 271–274.

Serrelli, E. (2015). Visualizing macroevolution: From adaptive landscapes to compositions of multiple spaces. In E. Serrelli & N. Gontier (Eds.), *Macroevolution: Explanation, interpretation and evidence* (Interdisciplinary evolution research series, pp. 113–162). Cham: Springer. doi:10.1007/978-3-319-15045-1_4.

Serrelli, E., & Gontier, N. (Eds.). (2015). *Macroevolution: Explanation, interpretation and evidence*. Springer. doi:10.1007/978-3-319-15045-1.

Shackleton, M., et al. (2009). Heterogeneity in cancer: Cancer stem cells versus clonal evolution. *Cell, 138*(5), 822–829.

Shamblott, M. J., et al. (1998). Derivation of pluripotent stem cells from cultured human primordial germ cells. *Proceedings of the National Academy of Sciences of the United States of America, 95*(23), 13726–13731.

Sharpless, N. E., & De Pinho, R. A. (2007). How stem cells age and why this makes us grow old. *Nature Reviews Molecular Cell Biology, 8*, 703–713.

Shih, C., & Weinberg, R. A. (1982). Isolation of a transforming sequence from a human bladder carcinoma cell line. *Cell, 29*, 161–169.

Shih, C., Shilo, B. Z., Goidfarb, M. P., Dannenberg, A., & Weinberg, R. A. (1979). Passage of phenotypes of chemically transformed ceils via transfection of DNA and chromatin. *Proceedings of the National Academy of Sciences of the United States of America, 76*, 5714–5718.

Silberstein, M. (2008). Reduction, Emergence and Explanation. In P. Machamer & M. Silberstein (Eds.), *The Blackwell guide to the philosophy of science* (pp. 80–107). Oxford: Blackwell Publishers Ltd.

Silva, S., Wiener, F., Klein, G., & Janz, S. (2005). Location of Myc, Igh, and Igk on Robertsonian fusion chromosomes is inconsequential for Myc translocations and plasmacytoma development in mice, but Rb(6.15)-carrying tumors prefer Igk-Myc inversions over translocations. *Genes, Chromosomes & Cancer, 42*(4), 416–426.

Skakkebék, N. E., Rajpert-De Meyts, E., Jørgensen, N., Carlsen, E., Petersen, P. M., Giwercman, A., Andersen, A. G., Jensen, T. K., Andersson, A. M., & Muller, J. (1998). Germ cell cancer and disorders of spermatogenesis: An environmental connection? *APMIS, 106*, 3–12.

Slaughter, D. P., Southwick, H. W., & Smejkal, W. (1953). Field cancerization in oral stratified squamous epithelium; clinical implications of multicentric origin. *Cancer, 6*, 963–968.

Sober, E. (2010). The multiple realizability argument against reductionism. In A. Rosenberg & R. Arp (Eds.), *Philosophy of biology: An anthology*. Hoboken: Wiley-Blackwell.

Sonnenschein, C., & Soto, A. M. (1999). *The society of cells: Cancer and control of cell proliferation*. New York: Springer.

Sonnenschein, C., & Soto, A. M. (2000). Somatic mutation theory of carcinogenesis: Why it should be dropped and replaced. *Molecular Carcinogenesis, 29*, 205–211.

Sonnenschein, C., & Soto, A. M. (2005). Are times a 'Changin' in carcinogenesis? *Endocrinology, 146*, 11–12.

Sonnenschein, C., & Soto, A. M. (2008). Theories of carcinogenesis: An emerging perspective. *Seminars in Cancer Biology, 18*, 372–377.

Sonnenschein, C., & Soto, A. M. (2011). Response to "In defense of the somatic mutation theory of cancer". *Bioessays, 2011*(33), 657–659.

Sonnenschein, C., & Soto, A. M. (2013a). Cancer genes: The vestigial remains of a fallen theory. In S. Krimsky & J. Gruber (Eds.), *Genetic explanations: Sense and nonsense* (pp. 81–93). Cambridge: Harvard University Press.

Sonnenschein, C., & Soto, A. M. (2013b). The aging of the 2000 and 2011 Hallmarks of Cancer reviews: A critique. *Journal of Biosciences, 38*, 1–13.

Soto, A. M., & Sonnenschein, C. (1999). *The society of cells: Cancer and control of cell proliferation*. New York: Springer.

Soto, A. M., & Sonnenschein, C. (2004). The somatic mutation theory of cancer: Growing problems with the paradigm? *BioEssays, 26*, 1097–1107.

Soto, A. M., & Sonnenschein, C. (2005). Emergentism as a default: Cancer as a problem of tissue organization. *Journal of Biosciences, 30*, 103–118.

Soto, A. M., & Sonnenschein, C. (2006a). Correcting an error. *BioEssays, 28*(2), 227–227.

Soto, A. M., & Sonnenschein, C. (2006b). Emergentism by default: A view from the bench. *Synthese, 151*(3), 361–376.

Soto, A. M., & Sonnenschein, C. (2010). Environmental causes of cancer: Endocrine disruptors as carcinogens. *Nature Reviews Endrocrinology, 6*(7), 363–370.

Soto, A. M., & Sonnenschein, C. (2011). The tissue organization field theory of cancer: A testable replacement for the somatic mutation theory. *Bioessays, 33*, 332–340.

Soto, A. M., Maffini, M. V., & Sonnenschein, C. (2008a). Neoplasia as development gone awry: The role of endocrine disruptors. *International Journal of Andrology, 31*(2), 288–293.

Soto, A. M., Sonnenschein, C., & Miquel, P. A. (2008b). On physicalism and downward causation in developmental and cancer biology. *Acta Biotheoretica, 56*(4), 257–274.

Spencer, S. L., Gerety, R. A., Pienta, K. J., & Forrest, S. (2006). Modeling somatic evolution in tumorigenesis. *PLoS Computational Biology, 2*, e108.

Spencer, V. A., Xu, R., & Bissell, M. J. (2007). Extracellular matrix, nuclear and chromatin structure, and gene expression in normal tissues and malignant tumors: A work in progress. *Advances in Cancer Research, 97*, 275–294.

Sporn, M. B. (1991). Carcinogenesis and cancer: Different perspectives on the same disease. *Cancer Research, 51*, 6215–6218.

Sporn, M. B. (2006). Dichotomies in cancer research: Some suggestions for a new synthesis. *National Clinical Practice Oncology, 3*, 364–373.

Steel, D. M., & Harris, H. (1989). The effect of antisense RNA to fibronectin on the malignancy of hybrids between melanoma cells and normal fibroblasts. *Journal of Cell Science, 93*, 515–524.

Stengers, I. (1997). *Cosmopolitiques VI*. Paris: La découverte.

Stephan, A. (1999). *Emergenz. Von der Unvorhersagbarket zur Selstorganisation*. Dresden: Dresden University Press.

Sternlicht, M. D., Lochter, A., Sympson, C. J., Huey, B., Rougier, J. P., Gray, J. W., Pinkel, D., Bissell, M. J., & Werb, Z. (1999). The stromal proteinase MMP3/stromelysin-1 promotes mammary carcinogenesis. *Cell, 98*, 137–146.

Strange, K. (2005). The end of "naïve reductionism": Rise of systems biology or renaissance of physiology? *American Journal of Physiology—Cell Physiology, 288*(5), C968–C974.

Strohman, R. C. (1997). The coming Kuhnian revolution in biology. *Nature Biotechnology, 15*, 194–200.

Tabin, C. J., Bradley, S. M., Bargmann, C. I., Weinberg, R. A., Papageorge, A. G., Scolnick, E. M., Dhar, R., Lowy, D. R., & Chang, E. H. (1982). Mechanism of activation of a human oncogene. *Nature, 300*, 143–149.

Tellegen, B. D. H. (1952, August). A general network theorem, with applications. *Philips Research Reports, 7*, 259–269.

Thiery, J. P. (2002). Epithelial-mesenchymal transitions in tumour progression. *Nature Reviews Cancer, 2*, 442–454.

Thomson, J. A., Itskovitz-Eldor, J., Shapiro, S. S., Waknitz, M. A., Swiergiel, J. J., Marshall, V. S., & Jones, J. M. (1998). Embryonic stem cell lines derived from human blastocysts. *Science, 282*, 1145–1147.

Toepfer, G. (2012). Teleology and its constitutive role for biology as the science of organized systems in nature. *Studies in History and Philosophy of Science, 43*, 113–119.

Tokunaga, M., Norman, J. E., Asano, M., Tokuoka, S., Ezaki, H., Nishimori, I., & Tsuji, Y. (1979). Malignant breast tumors among atomic bomb survivors, Hiroshima and Nagasaki, 1950–74. *Journal of the National Cancer Institute, 62*, 1347–1359.

Urbani Ulivi, L. (Ed.). (2011a). *Strutture di mondo. Il pensiero sistemico come specchio di una realtà complessa*. Bologna: Il Mulino.

Urbani Ulivi, L. (2011b). La struttura dell'umano. Linee di un'antropologia sistemica. In *Strutture di mondo. Il pensiero sistemico come specchio di una realtà complessa*. Bologna: Il Mulino.

Ushijima, T. (2007). Epigenetic field for cancerization. *Journal of Biochemistry and Molecular Biology, 40*, 142–150.
van Gent, D. C., Hoeijmakers, J. H., & Kanaar, R. (2001). Chromosomal stability and the DNA double-stranded break connection. *Nature Reviews Genetics, 2*(3), 196–206.
Varmus, H. E., & Weinberg, R. A. (1992). *Genes and the biology of cancer.* New York: Scientific American Library.
Vaux, D. L. (2011). In defense of the somatic mutation theory of cancer. *Bioessays, 33*(5), 341–343.
Vaux, D. L., & Korsmeyer, S. J. (1999). Cell death in development. *Cell, 96*, 245–254.
Vermeulen, L., Sprick, M. R., Kemper, K., Stassi, G., & Medema, J. P. (2008). Cancer stem cells – old concepts, new insights. *Cell Death and Differentiation, 15*, 947–958.
Vescovi, A. L., Galli, R., & Reynolds, B. A. (2006). Brain tumour stem cells. *Nature Reviews Cancer, 6*, 425–436.
Vineis, P., Schatzkin, A., & Potter, J. D. (2010). Models of carcinogenesis: An overview. *Carcinogenesis, 31*, 1703–1709.
Vitiello, G. (2001). *My double Unveiled, advances in consciousness research.* Amsterdam/Philadelphia: John Benjamin Publishing Company.
Vogelstein, B., & Kinzler, K. W. (2004). Cancer genes and the pathways they control. *Nature Medicine, 10*, 789–799.
Von Bertalanffy, L. (1968). *General system theory.* New York: Braziller.
Waddington, C. H. (1935). Cancer and the theory of organizers. *Nature, 135*, 606–608.
Waddington, C. H. (1977). *Tools for thought* (p. 145). London: Jonathan Cape Thirty Bedford Square.
Wakefield, L. M., & Sporn, M. B. (1990). Suppression of carcinogenesis: A role for TGF-/3 and related molecules in prevention of cancer. In *Tumor suppressor genes*. New York: G. Klein. Marcel Dekker, Inc.
Waliszewski, P., Molski, M., & Konarski, J. (2001). On the relationship between fractal geometry of space and time in which a system of interacting cells exists and dynamics of gene expression. *Acta Biochimica Polonica, 48*(1), 209–220.
Wang, W., et al. (2002). Single cell behavior in metastatic primary mammary tumors correlated with gene expression patterns revealed by molecular profiling. *Cancer Research, 62*(21), 6278–88.
Wasan, H. S., et al. (1998). APC in the regulation of intestinal crypt fission. *The Journal of Pathology, 185*(3), 246–255.
Watson, J. D., & Crick, F. H. (1953a). Molecular structure of nucleic acids; A structure for deoxyribose nucleic acid. *Nature, 171*, 737–738.
Watson, J. D., & Crick, F. H. (1953b). Genetical implications of the structure of deoxyribonucleic acid. *Nature, 171*, 964–967.
Weaver, V. M., et al. (1997). Reversion of the malignant phenotype of human breast cells in three-dimensional culture and In vivo by integrin blocking antibodies. *The Journal of Cell Biology, 137*(1), 231–245.
Weinberg, R. A. (1988). The genetic origins of human cancer. *Cancer, 61*, 1963–1968.
Weinberg, R. A. (Ed.). (1989). *Oncogenes and the molecular origins of cancer*. New York: Cold Spring Harbor Laboratory Press.
Weinberg, R. A. (1998). *One renegade cell: How cancer begins*. New York: Basic Books.
Weinberg, R. A. (2006). *The biology of cancer*. London: Garland Science.
Weinberg, R. A. (2008). Mechanisms of malignant progression. *Carcinogenesis, 29*, 1092–1095.
Weinberg, R. A. (2014). Coming full circle – From endless complexity to simplicity and back again. *Cell, 157*(1), 267–271.
Westerhoff, H. V., & Kell, D. B. (2007). The methodologies of systems biology. In F. C. Boogerd, F. J. Bruggeman, J. H. S. Hofmeyr, & H. V. Westerhoff (Eds.), *Systems biology: Philosophical foundations*. Amsterdam: Elsevier.

Wijnhoven, S. W., Hoogervorst, E. M., de Waard, H., van der Horst, G. T., & van Steeg, H. (2007). Tissue specific mutagenic and carcinogenic responses in NER defective mouse models. *Mutation Research, 614*, 77–94.

Wimsatt, W. C. (2007). *Re-engineering philosophy for limited beings: Piecewise approximations to reality, Ch 6*. Cambridge: Harvard University Press.

Winther, R. G. (2011). Part-whole science. *Synthese, 178*, 397–427.

Wolfe, C. T. (2012). Chance between holism and reductionism: Tensions in the conceptualization of Life. *Progress in Biophysics and Molecular Biology, 110*, 113–120.

Woodward, J. (2003). *Making things happen: A theory of causal explanation*. Oxford: Oxford University Press.

Woodward, J. (2006). Sensitive and insensitive causation. *The Philosophical Review, 115*, 1–50.

Woodward, J. (2010). Causation in biology: Stability, specificity, and the choice of levels of explanation. *Biology and Philosophy, 25*, 287–318.

Woodward, J. (2011). Scientific explanation. In E. N. Zalta (Ed.), *The Stanford encyclopedia of philosophy*.

Wright, L. (1973). Functions. *Philosophical Review, 92*, 139–168.

Xu, R., Boudreau, A., & Bissell, M. J. (2009a). Tissue architecture and function: Dynamic reciprocity via extra- and intra-cellular matrices. *Cancer and Metastasis Reviews, 28*, 167–176.

Xu, R., Nelson, C. M., Muschler, J. L., Veiseh, M., Vonderhaar, B. K., & Bissell, M. J. (2009b). Sustained activation of STAT5 is essential for chromatin remodeling and maintenance of mammary specific function. *Journal of Cell Biology, 184*, 57–66.

Yablo, S. (1992). Mental causation. *Philosophical Review, 101*, 245–280.

Yachida, S., Jones, S., Bozic, I., Antal, T., Leary, R., Fu, B., Kamiyama, M., Hruban, R. H., Eshleman, J. R., Nowak, M. A., Velculescu, V. E., Kinzler, K. W., Vogelstein, B., & Iacobuzio-Donahue, C. A. (2010). Distant metastasis occurs late during the genetic evolution of pancreatic cancer. *Nature, 467*(7319), 1114–1117.

Yale, S. H., Vasudeva, S., Mazza, J. J., Rolak, L., Arrowood, J., Stichert, S., & Stratman, E. S. (2005). Disorders of flushing. *Comprehensive Therapy, 31*, 59–71.

Ye, C. J., Liu, G., Bremer, S. W., & Heng, H. H. (2007). The dynamics of cancer chromosomes and genomes. *Cytogenetic and Genome Research, 118*, 237–246.

Ye, C. J., Stevens, J. B., Liu, G., Bremer, S. W., Jaiswal, A. S., Ye, K. J., Lin, M. F., Lawrenson, L., Lancaster, W. D., Kurkinen, M., Liao, J. D., Gairola, C. G., Shekhar, M. P., Narayan, S., Miller, F. R., & Heng, H. H. (2009). Genome based cell population heterogeneity promotes tumorigenicity: The evolutionary mechanism of cancer. *Journal of Cellular Physiology, 219*, 288–300.

Zhang, Q., et al. (2001). Nesprins: A novel family of spectrin-repeat-containing proteins that localize to the nuclear membrane in multiple tissues. *Journal of Cell Science, 114*(Pt 24), 4485–4498.

MIX
Papier aus verantwortungsvollen Quellen
Paper from responsible sources
FSC® C105338

If you have any concerns about our products,
you can contact us on
ProductSafety@springernature.com

In case Publisher is established outside the EU,
the EU authorized representative is:
Springer Nature Customer Service Center GmbH
Europaplatz 3, 69115 Heidelberg, Germany

Printed by Libri Plureos GmbH
in Hamburg, Germany